# Advances in Experimental Medicine and Biology

Volume 902

Advances in Experimental Medicine and Biology presents multidisciplinary and dynamic findings in the broad fields of experimental medicine and biology. The wide variety in topics it presents offers readers multiple perspectives on a variety of disciplines including neuroscience, microbiology, immunology, biochemistry, biomedical engineering and cancer research. Advances in Experimental Medicine and Biology has been publishing exceptional works in the field for over 30 years and is indexed in Medline, Scopus, EMBASE, BIOSIS, Biological Abstracts, CSA, Biological Sciences and Living Resources (ASFA-1), and Biological Sciences. The series also provides scientists with up to date information on emerging topics and techniques.

More information about this series at http://www.springer.com/series/5584

Andreas Schwiertz
Editor

# Microbiota of the Human Body

## Implications in Health and Disease

Springer

*Editor*
Andreas Schwiertz
Institute of Microecology
Herborn, Germany

ISSN 0065-2598 ISSN 2214-8019 (electronic)
Advances in Experimental Medicine and Biology
ISBN 978-3-319-31246-0 ISBN 978-3-319-31248-4 (eBook)
DOI 10.1007/978-3-319-31248-4

Library of Congress Control Number: 2016939968



Printed on acid-free paper

This Springer imprint is published by Springer Nature
The registered company is Springer International Publishing AG Switzerland

# Preface

Microbes can now be found in nearly every niche the human body offers. However, the complexity of the microbiota of a given site depends on the particular environmental condition thereof. Only microbes that are able to grow under these conditions will prevail. Recent publications imply that the microorganisms do not only have multiple critical consequences for host physiological processes, such as postnatal development, immunomodulation and energy supply, but also effects on neurodevelopment, behaviour and cognition.

Within this book we will focus on the techniques behind these developments, epigenomics and on the various parts of the human body, which are inhabited by microorganisms, such as the mouth, the gut, the skin and the vagina. In addition, chapters are dedicated to the possible manipulations of the microbiota by probiotics, prebiotics and faecal transplantation.

I would like to express my gratitude to all chapters' authors for their contribution to this book and hope that it will be appreciated by readers as well as it occurred to me as an editor.

Herborn, Germany Andreas Schwiertz

# Contents

# A Short Definition of Terms

# 1

Andreas Schwiertz and Volker Rusch

**Keywords**
Definition • Microbiota • Microbiome

We humans are colonized by myriads of microorganisms in various parts of the body, such as the skin, the mouth, the vagina and the gastrointestinal tract. Even the lung and other hitherto thought to be sterile parts, as the placenta, are now considered to be colonized. Furthermore, our microbiota is not only comprised of bacteria, but also of archaea and eukaryotes such as protozoa, fungi and nematodes. Even viruses, collectively termed the virome, can be found in the microbiota (Virgin 2014). It has been estimated that the human-associated microbiota, consists of at least 40,000 bacterial strains in 1800 genera (Luckey 1972; Frank and Pace 2008; Forsythe and Kunze 2013), which collectively harbor at least 9.9 million non-human genes (Li et al. 2014). They encode for approximately 500 times the human protein-coding genes which are currently annotated (http://www.ensembl.org). The estimated mass of the microbiota (1–2 kg in an adult body (Forsythe and Kunze 2013)) is comparable to the weight of the adult human brain (ca. 1.5 kg, Parent and Carpenter 1996).

As of today our knowledge on the human microbiota is due to the fast evolution of sequencing. On the 14th of April 2003 the completion of the human genome sequencing process was announced and in 2004 the quality assessment of the human genome sequence finally published. Since then huge efforts have been undertaken to sequence other important genomes like that of the rat (*rattus norvegicus*), the honey bee (*apis mellifera*) and even the Neanderthal. In 2008 the national institute of health decided to fund the Human Microbiome Project (HMP). Goal was the "*characterization of the human* ***microbiome*** *and analysis of its role in human health and disease*" (http://hmpdacc.org/) (Turnbaugh et al. 2007; Human Microbiome Project Consortium 2012). In parallel the MetaHIT project financed by the European Commission under the 7th FP program was launched. Its aim was to "*establish associations between the genes of the human intestinal* ***microbiota*** *and our health and disease*" (http://www.metahit.eu/) (Qin et al. 2010).

A. Schwiertz (✉)
Institute of Microecology,
Auf den Lueppen 8, 35745 Herborn, Germany
e-mail: andreas.schwiertz@mikrooek.de

V. Rusch
Institute of Integrative Biology,
35745 Herborn, Germany
e-mail: rusch@rusch-chateau.ch


A. Schwiertz (ed.), *Microbiota of the Human Body*, Advances in Experimental Medicine and Biology 902, DOI 10.1007/978-3-319-31248-4_1

Interestingly, as seen in the two mentioned projects the terms microbiome and microbiota are used analogical even if their scientific definition is dissimilar. The term "**microbiota**" describes the total collection of organisms of a geographic region or a time period. Searching Google scholar and Pubmed the first appearance of the term microbiota in connection with bacteria was a patent application by Alexander Goetz from 1945 (Goetz 1950). In the context of human health the term microbiota was first used to describe the gingival crevice (Socransky et al. 1953), while it did not appear before 1966 for the description of the biggest accumulation of bacteria within the human body the gastrointestinal microbiota (Dubos 1966). The term "microbiome" was originally used to refer to the collection of the genomes of the microbes in a particular ecosystem and

**Fig. 1.1** Definitions of terms

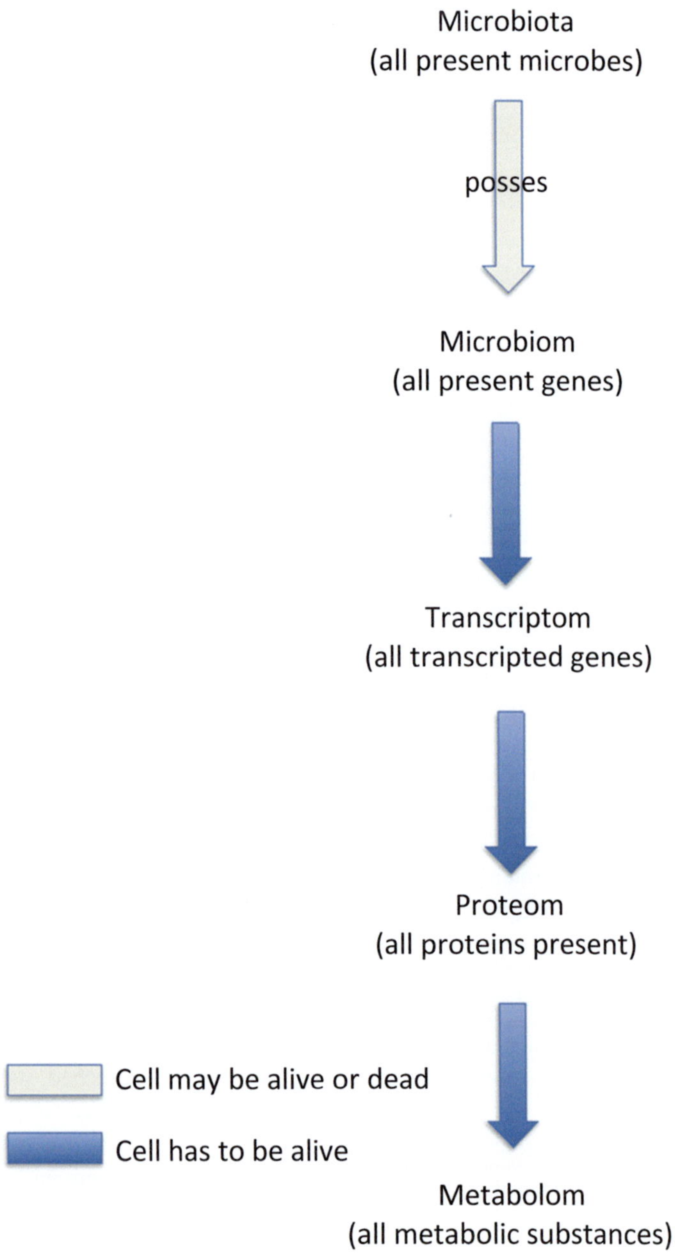

termed by Nobel laureate Joshua Lederberg (1925–2008) (Hooper and Gordon 2001). Therefore, the term microbiota would be correct in the case of 16S rRNA studies and the term microbiome in genome studies. To study the microbiota of a given habitat and the therein present genes the microbes of interest may be dead or alive. However, the applied techniques do not allow for the discrimination between a living or a dead cell, in contrast to the determination of the transcriptom, proteome or the metabolom of a habitat (Fig. 1.1).

Only the determination of the latter three will allow us insights on the implication and importance of a specific microbe in a habitat and not only an ordinary number.

As we humans are not only determined by our genes, but by the transcribed proteins, so is not the sole microbe of importance, but its liaison to other microbes and us.

As Louis Pasteur once stated:"*Le microbe, c'est rien, le milieu, c'est tout! – The microbe is noting, it's the environment*".

## References

Dubos R (1966) The microbiota of the gastrointestinal tract. Gastroenterology 51:868–874

Forsythe P, Kunze WA (2013) Voices from within: gut microbes and the CNS. Cell Mol Life Sci 70:55–69. doi:10.1007/s00018-012-1028-z

Frank DN, Pace NR (2008) Gastrointestinal microbiology enters the metagenomics era. Curr Opin Gastroenterol 24:4–10. doi:10.1097/MOG.0b013e3282f2b0e8

Goetz A (1950) Method for producing a microbicidal composition of matter. Google Patents

Hooper LV, Gordon JI (2001) Commensal host-bacterial relationships in the gut. Science 292:1115–1118

Li J, Jia H, Cai X, Zhong H, Feng Q, Sunagawa S et al (2014) An integrated catalog of reference genes in the human gut microbiome. Nat Biotechnol 32:834–841. doi:10.1038/nbt.2942

Luckey TD (1972) Introduction to intestinal microecology. Am J Clin Nutr 25:1292–1294

Parent A, Carpenter MB (1996) Carpenter's human neuroanatomy. Williams & Wilkins, Baltimore

Qin J, Li R, Raes J, Arumugam M, Burgdorf KS, Manichanh C et al (2010) A human gut microbial gene catalogue established by metagenomic sequencing. Nature 464:59–65. doi:10.1038/nature08821

Socransky SS, Gibbons RJ, Dale AC (1953) The microbiota of the gingival crevice area of man. I. Total microscopic and viable counts of specific microorganisms. J Arch Oral Biol 8:275–280

The human microbiome project consortium (2012) Structure, function and diversity of the healthy human microbiome. Nature 486:207–214

Turnbaugh PJ, Ley RE, Hamady M, Fraser-Liggett CM, Knight R, Gordon JI (2007) The human microbiome project. Nature 449:804–810. doi:10.1038/nature06244

Virgin HW (2014) The virome in mammalian physiology and disease. Cell 157:142–150. doi:10.1016/j.cell.2014.02.032

# Studying the Human Microbiota

# 2

Alan W. Walker

**Abstract**

There are a range of methodologies available to study the human microbiota, ranging from traditional approaches such as culturing through to state-of-the-art developments in next generation DNA sequencing technologies. The advent of molecular techniques in particular has opened up tremendous new avenues for research, and has galvanised interest in the study of our microbial inhabitants. Given the dazzling array of available options, however, it is important to understand the inherent advantages and limitations of each technique so that the best approach can be employed to address the particular research objective. In this chapter we cover some of the most widely used current techniques in human microbiota research and highlight the particular strengths and caveats associated with each approach.

**Keywords**

Microbiota • Techniques • Sequencing • PCR • FISH • Stable isotope • Metabolomics • Proteomics

## 2.1 Introduction

The Nobel prize winning biologist Sydney Brenner once remarked that "progress in science results from new technologies, new discoveries and new ideas, probably in that order" (Robertson 1980) and this sentiment has undoubtedly been well exemplified in the field of microbiota research. Study of the human microbiota can be traced back to Antonie van Leeuwenhoek's late Seventeenth Century description of "animalcules" in scrapings from the human mouth (Porter 1976), a discovery that was made possible by van Leeuwenhoek's ground-breaking work with microscopes. From the pioneering endeavours of Cohn, Pasteur, Koch and others in the Nineteenth Century, through to developments in anaerobic microbiology and molecular biology in the

A.W. Walker (✉)
Microbiology Group, Rowett Institute of Nutrition and Health, University of Aberdeen, Foresterhill, Aberdeen AB25 2ZD, UK
e-mail: alan.walker@abdn.ac.uk





A. Schwiertz (ed.), *Microbiota of the Human Body*, Advances in Experimental Medicine and Biology 902, DOI 10.1007/978-3-319-31248-4_2

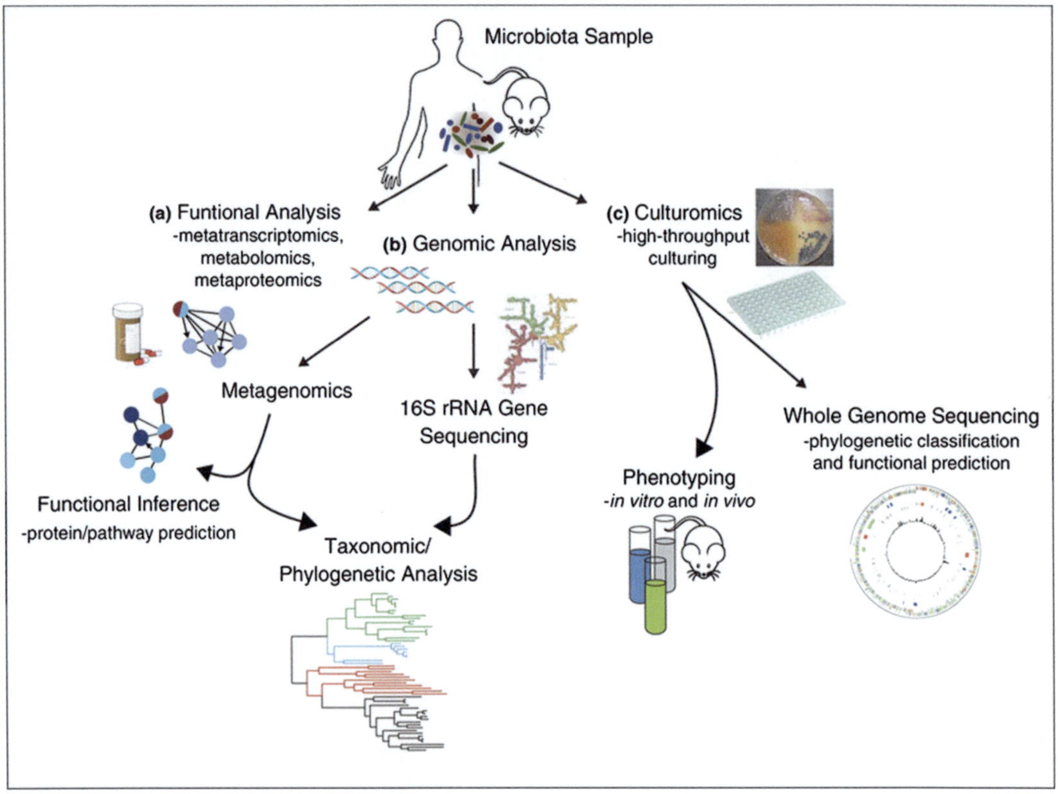


**Fig. 2.1 Overview of some of the most common techniques used to study the human microbiota**
**(a)** The functional activities of the microbiota can be studied by monitoring transcription (using RNA-seq/metatranscriptomics), protein production (metaproteomics) or metabolite production (metabolomics). **(b)** DNA sequence-based techniques are used to determine the composition of the microbiota (e.g. 16S rRNA gene surveys) and the functional encoding capabilities of the microbiome (shotgun metagenomics). **(c)** Culture remains highly relevant as cultured organisms can be studied in depth in the laboratory or in animal hosts. Recently, the term "culturomics" has been applied to high-throughput culturing of microbes in multi-welled plates containing highly nutritious growth media. Cultured organisms can also have their genomes sequenced, providing further information about their potential activities *in vivo*. These techniques can be used in combination to generate more comprehensive understandings of the human microbiota. Reprinted in unmodified form from: Pham and Lawley (2014) (Pham and Lawley 2014) under Creative Commons Attribution (CC BY) license

second half of the Twentieth Century, and the Twenty-first Century's own breakthroughs in genomics and DNA sequencing technologies (McPherson 2014), subsequent developments in the field of microbiota research have been similarly driven by successive waves of technological and methodological advances. As a result, today's microbiota researcher has the benefit of a staggering array of tools at their disposal (Fig. 2.1). This chapter gives a broad overview of the many techniques that are now available, and attempts to describe the inherent advantages and limitations of each of these techniques.

## 2.2 Classical Microbiological Methods

### 2.2.1 Culture

For well over a 100 years microbiologists have used the classical approaches of cultivating microbes in the laboratory, isolating individual colonies and then studying these isolated strains in order to describe their phenotypic characteristics and metabolic capabilities (see Lagier et al. (2015a) for a recent overview of the techniques used). As a result of these extensive efforts, it has been estimated that over 1000 distinct microbial species have been cultured from the human gastrointestinal tract alone (McPherson 2014), and characterisation of microbes and gene function discovery in the laboratory remains the bedrock upon which many of the more modern molecular techniques that will be described in later sections of this chapter rely upon. A further advantage of having a strain in culture is that it allows potential exploitation for therapeutic purposes should it turn out to have beneficial properties (Walker et al. 2014).

The simplest form of microbial cultivation is to incubate samples or individual strains in batch culture in nutritious or selective growth media. Batch culture studies allow selective enrichment of bacterial groups of interest, comparisons to be made between growth rates and metabolite production on different substrates, and interactions between specific species to be observed and measured (Belenguer et al. 2006). Many microbial inhabitants of humans are obligately anaerobic and therefore exquisitely sensitive to oxygen. As a result, some species can be killed by even very brief exposure to air (Flint et al. 2007), making them much more difficult to grow. To permit laboratory cultivation of these species, culturing must therefore be carried out under strictly anaerobic conditions, for example by using anaerobic cabinets or Hungate roll tubes (Eller et al. 1971). Cultivation of particularly fastidious gut species can also be enhanced by using media containing rumen fluid, filtered stool extracts, or mixtures of short chain fatty acids, which can be utilised by some gut bacteria as growth substrates (Duncan et al. 2002; Lagier et al. 2015b).

A limitation of batch culture is that results can only be obtained over relatively short periods of time before the supply of nutrients in the growth medium is exhausted or toxic by-products accumulate and lead to cessation of microbial growth (Ferenci 1999). A further, and key, disadvantage to using culture is that it is highly labour intensive, and a range of complex growth media are typically required to recover as wide a diversity of organisms from a sample as possible. It is also known that many of the microbial species that inhabit the human body have yet to be grown in the laboratory (Rajilic-Stojanovic et al. 2007). This problem is particularly acute for bodily sites such as the colon, where the majority of the constituent bacteria are strict anaerobes. As such, culture alone cannot address the sheer complexity of the human microbiota.

Nonetheless, there are many reasons to be optimistic that cultured coverage of the human microbiota can be greatly improved. DNA-sequence based surveys of the gut microbiota, for example, commonly show that many of the most abundant sequences map to cultured species, and that it is the rarer sequences that are less likely to be derived from a cultured isolate (Walker et al. 2014). This suggests that it is insufficient culturing effort rather than an inherent "unculturability" that is the main barrier to successful novel isolations. Furthermore, unlike environments such as soil, which can harbour very slow growing microbes, bacteria living in the human body are often provided with relatively stable environmental conditions, and a generally reliable supply of growth nutrients, and must therefore be capable of multiplying quickly or else face being rapidly outcompeted. Provided the correct conditions can be supplied in artificial growth media it can be assumed therefore that these species will be relatively more amenable to culture. Indeed, novel species continue to be regularly isolated from the human microbiota, and there have been some impressive recent examples of successful high-throughput culturing programmes (Lagier et al. 2015b; Goodman et al. 2011). Such efforts have been dubbed "culturomics", and have contributed to a reinvigorated interest in the use of culture-based techniques to better characterise the human microbiota. Information gleaned from modern genomics methods can also be used to design improved culture media that support the growth of previously uncultivated species (Bomar et al. 2011).

### 2.2.2 Continuous Culture

A more sophisticated method to cultivate microbes in the laboratory is the use of continuous culture model systems such as fermentors

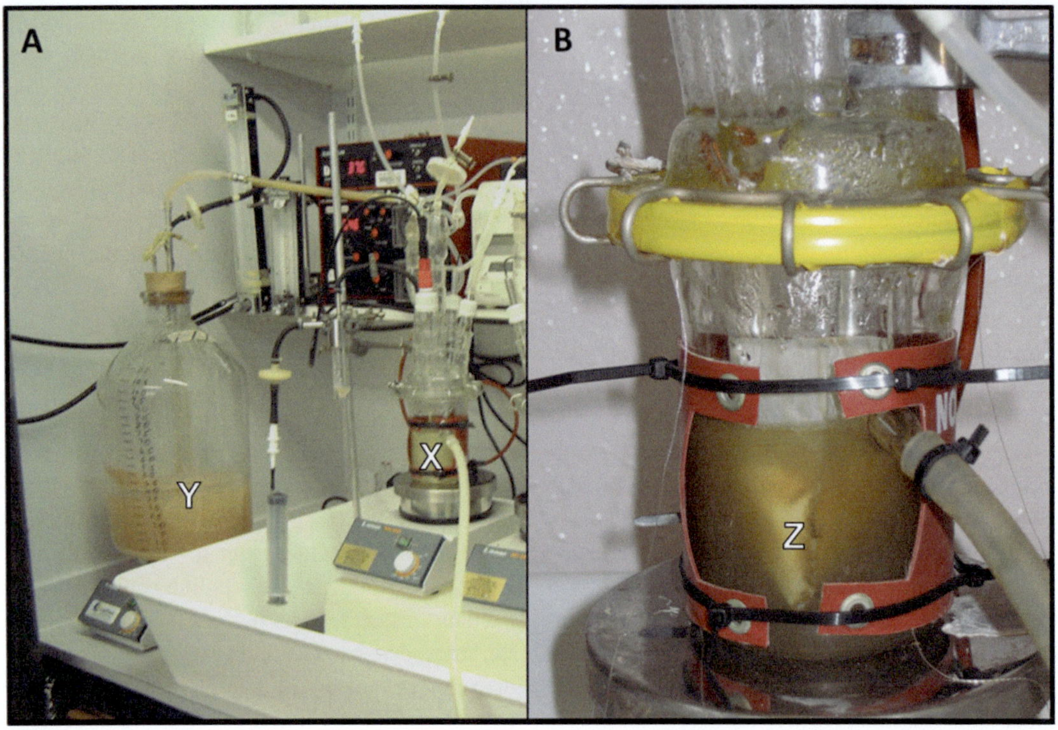


**Fig. 2.2** **Continuous culture fermentor system** Fermentors are continuous culture model systems, which allow long term cultivation of microbes. **(A)** An example of a single vessel fermentor system (the culture vessel is labelled with "X"), inoculated with human faeces and fed a constant supply of nutritious growth medium (labelled "Y"). The contents of the culture vessel are gassed with $CO_2$ or $N_2$ to ensure that they remain anaerobic, and can be maintained at defined pH and temperatures, which are constantly monitored. **(B)** a modified fermentor vessel, incorporating a nylon bag containing insoluble particulate substrates (labelled "Z"), developed to identify fibre-degrading gut bacteria

(Fig. 2.2). In contrast to the batch approach, continuous culture is carried out in an open system, which is continually supplied at one end with fresh growth medium/nutrients, and overflow is allowed to drain from the vessel at the other end, diluting out toxic metabolic by-products and dead cells. Systems such as these reach a "steady state" equilibrium, allowing the researcher to exert an enhanced level of control over prevailing environmental conditions within the culture vessel, and can therefore be run over relatively long time periods (Miller and Wolin 1981). These sort of systems have been commonly used to study colonic microbes, and a number of research groups have made fermentors more advanced by incorporating distinct sequential stages, which aim to mimic the sort of environmental changes microbes are exposed to as they pass along the length of the gastrointestinal tract (Van den Abbeele et al. 2010). While these model systems are an advance over simple batch culture it should be noted, however, that they still have important limitations. For example, they lack an immune system, and metabolites such as short chain fatty acids (SCFAs) produced by the bacteria are not absorbed, meaning results may not necessarily be directly translatable to the situation *in vivo*.

### 2.2.3 Animal Models

Microbes of interest can also be cultivated and maintained in animal models. Until relatively recently, for example, the only way to grow segmented filamentous bacteria, which have been shown to have important pro-inflammatory effects in mice, was in animal models (Klaasen et al. 1991). One disadvantage of using animal models is that, while the microbiota composition at the phylum level generally appears to be similar between humans and other animals, at the species and strain level there is considerable divergence, likely to due to underlying differences in host anatomy/physiology, and dietary regimes (Nguyen et al. 2015). However, recent work has shown that it may be possible to mitigate this issue somewhat as a significant proportion of human-associated bacterial species appear to be able to successfully colonise the intestines of animal models following faecal microbiota transfer (Ellekilde et al. 2014). Germ-free, or gnotobiotic, mice are another appealing option as these mice can be specifically inoculated with microbial strains of interest (Goodman et al. 2011; Seedorf et al. 2014). This permits a more reductionist approach to study host-microbe and microbe-microbe interactions, separated from the potentially perplexing background complexity of the wider microbiota. A further particular advantage of using mouse models is that extensive genotyping analyses have been carried out, and there are a range of knockout mouse lines available to allow the study of interactions between specific host genetic components and the microbiota (Kostic et al. 2013).

There are, however, a number of important limitations to using animal models, particularly rodent models. For example, co-housing, and the practice of coprophagy, generally leads to rapid transfer of microbiota between cage mates, and this can confound results by being a stronger determinant of intestinal microbiota composition than either host genotype or experimental variables (Lees et al. 2014; Ericsson et al. 2015). Furthermore, recent work has indicated that rodents who are handled by male experimenters are likely to be more stressed than those handled by females (Sorge et al. 2014), and it is possible that stress may impact microbiota structure (Cryan and Dinan 2012). Finally, emerging evidence suggests that host diet may have a greater impact on microbiota structure and composition in rodents than in humans (explaining around 60 % of variance vs 10 % respectively), raising concerns as to whether or not rodent models are most appropriate for studies investigating links between the microbiota and, for example, diet-dependent diseases such as obesity (Salonen et al. 2014). A recent review by Nguyen et al (2015) extensively documents the inherent advantages and disadvantages of using mouse models, and discusses the translatability of findings in mice to humans.

## 2.3 Sequence-Based Approaches

While culture remains an important tool, human microbiota research has been completely revolutionised over the last decade by molecular methods, and in particular by the falling costs and vastly increased throughput of DNA sequencing technologies (Fig. 2.3). This rapidly moving, and highly innovative, field continues to produce exciting and novel technologies, with the latest generation of sequencing machines capable of generating data at a depth of billions of individual sequence reads (Illumina HiSeq), or at comparatively long read lengths (PacBio), or even via miniaturised devices that can be plugged into the USB port of a laptop (Oxford Nanopore's MinION) (Reuter et al. 2015).

The key advantage to sequence-based approaches is that, by circumventing the requirement to grow microorganisms in the laboratory, they generally give much more comprehensive overviews of the species present in a sample. They are also typically far less labour intensive than classical microbiological techniques, and as a result it is now possible to carry out experiments at a scale that would have been unthinkable just a decade ago. Indeed, recent global research initiatives such as the Human Microbiome Project (HMP) and MetaHIT, for example, have taken advantage of these new sequencing technologies to produce staggering amounts of freely available data (Human Microbiome Project Consortium 2012a; Li et al. 2014). There are a number of ways in which the power of DNA sequencing can be used to study the human microbiota, which are detailed in the following text. In addition, Table 2.1 summarises

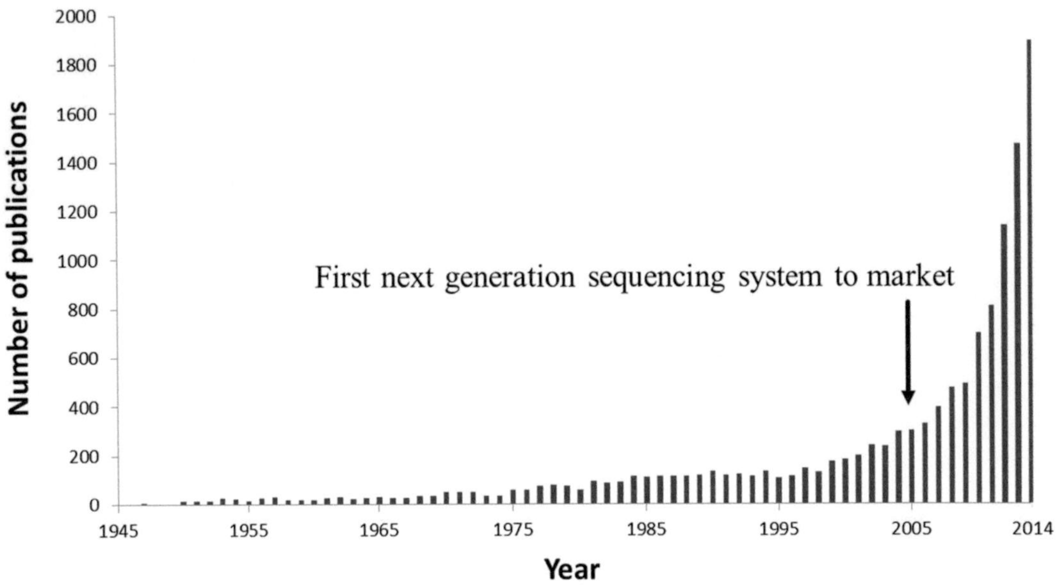


**Fig. 2.3 DNA sequencing approaches have revolutionised microbiota research**
Chart showing the meteoric rise in publications mentioning the gut microbiota since the advent and market release of next generation sequencing platforms such as 454 pyrosequencing and Illumina. Data collected by searching Pubmed (search date Dec. 1st, 2014) for the terms "gut flora" OR "gut microflora" OR "gut microbiota" OR "gut microbiome" OR "intestinal flora" OR "intestinal microflora" OR "intestinal microbiota" OR "intestinal microbiome" OR "colonic flora" OR "colonic microflora" OR "colonic microbiota" OR "colonic microbiome"

the most common uses, and outlines the inherent advantages and limitations of each approach.

### 2.3.1 Marker Gene Surveys

One common sequence-based approach is to carry out surveys of universal marker genes, which provide a broad census of the microbial species present within a sample. While these sort of surveys have been carried out since the 1980s recent developments in next generation sequencing technologies mean it is possible to survey microbial communities at previously unimaginable depth and scales (Tringe and Hugenholtz 2008; Caporaso et al. 2011). The most widely used universal marker genes are the small subunit ribosomal RNA (SSU rRNA) genes (16S rRNA gene for bacteria and archaea, 18S rRNA gene for eukaryotes). Within these genes there are regions of DNA sequence that are highly conserved, but there are also other regions that are more variable, and which are unique to certain microbial groups or genera (Woese and Fox 1977). Following DNA extraction from the human tissue sample, the SSU rRNA genes are typically PCR-amplified using primers targeted towards highly conserved regions of the gene. The aim here is to generate a mixed pool of PCR amplicons that are derived from as many of the bacterial species present in the original sample as possible, which are then sequenced *en masse*. Typically, the resulting data is then clustered by sequence similarity into Operational Taxonomic Units (OTUs), with the assumption being that these OTUs will be a reasonable approximation of the underlying species content of a given sample. It should be noted though that, due to the wide variation in 16S rRNA gene operon copy numbers between individual strains, results are not truly quantitative (Vetrovsky and Baldrian 2013). Furthermore, the chosen OTU sequence similarity threshold is both artificial and subjective and will not be able to accurately capture diversity correctly across the full range of genera present in a sample. Nonetheless, when the full

**Table 2.1** Comparison between different sequence-based approaches used to study the human microbiota

| Method | Advantages | Limitations |
|---|---|---|
| Species profiling via marker gene surveys (e.g. 16S rRNA gene) | Provides overview of species present in a sample | Relatively insensitive – can be impossible to derive species-level classifications for some genera |
| | Much cheaper than other sequencing methods | Only provides information on community composition, does not provide direct functional capability data |
| | Analysis requires less computational power | Single marker genes such as 16S rRNA genes typically only describe the bacterial/archaeal fraction of microbial communities. Does not describe viruses, fungi etc. that may also be present. |
| | Larger sample sets increase statistical power | Results can be heavily impacted by sampling, storage, PCR and DNA extraction biases |
| | Broad functional capabilities can often be inferred from 16S rRNA gene sequences by comparing to closely related isolates with fully sequenced genomes | 16S rRNA gene is usually multi-copy, and the number of copies is variable between species, meaning results are not truly quantitative |
| | | Usually does not discriminate between active and inactive/dead cells |
| Whole genome sequencing | Provides information on the complete coding potential of an organism | Usually requires that the organism be cultivated prior to sequencing the genome |
| | Draft bacterial genomes can now be generated very quickly and cheaply | Modern, short read, sequencing technologies will typically generate draft, not complete, genomes |
| | Data generated can be used for epidemiological purposes, e.g. for strain typing | Many constituent genes will be of unknown function |
| Metagenomics | Allows simultaneous profiling of both the functional capabilities and species composition of microbial communities | Can require very deep sequencing to achieve reasonable genome coverage, making it comparatively expensive |
| | Can simultaneously obtain genomic data of bacterial, archaeal, eukaryotic and viral origin | Often limited to small numbers of samples, which reduces statistical power |
| | Complete genomes of constituent species, including uncultured organisms, can be assembled | Data analysis may require large computational resources. |
| | No PCR bias | Assembling genomes can be challenging |
| | | Gaps in reference databases mean that large proportions of the genomic data are of unknown function |
| | | Biases introduced during sampling, storage and DNA extraction can impact results |
| | | Usually does not discriminate between active and inactive/dead cells |

(continued)

**Table 2.1** (continued)

| Method | Advantages | Limitations |
|---|---|---|
| Single-cell genomics | Provides genomic data from uncultured species | Isolating single cells typically requires access to expensive equipment (e.g. flow cytometry, micromanipulators) |
| | Allows placement of genomic data in a phylogenetic context | Genome amplification step introduces biases, making complete genome assembly challenging |
| | Data generated from uncultured species can improve reference databases for metagenomic analyses. | Sensitivity of the genome amplification step means that contamination is a constant concern and must be mitigated against |
| | | Biases introduced during sampling, storage and DNA extraction can impact results |
| | | Cannot discriminate between active and inactive/dead cells |
| Metatranscriptomics | Gives data on the functional activity of microbial communities | Short half-life of mRNA is an important limitation; sample selection and preservation are key concerns |
| | Focusses on the active members of the microbiota, results not as impacted by dead/inactive cells as other sequencing methods | Can be technically challenging, often need to deplete the far more abundant rRNA before sequencing mRNA |
| | Can often attribute source organisms to transcripts | Gaps in reference databases mean that large proportions of the genomic data are of unknown function |
| | | Biases introduced during sampling, storage and RNA extraction can impact results |

1500 bp sequence of the 16S rRNA gene is available, clustering into OTUs with 98.7–99 % sequence similarity appears to best fit species-level designations derived from culture work (Stackebrandt and Ebers 2011). However, as next generation sequencing technologies typically generate comparatively short read lengths, which are focussed on hyper-variable regions of the gene, slightly less stringent clustering is required and it is now most common to cluster OTUs with 97 % sequence similarity (Schloss and Westcott 2011).

Regardless of sequence similarity used, OTUs can be mapped against comprehensive reference databases such as SILVA, RDP, EzTaxon and Greengenes in order to assign taxonomic classifications to them (Quast et al. 2013; Cole et al. 2014; Chun et al. 2007; DeSantis et al. 2006). This provides information about which taxa were present in the original sample, and allows the researcher to monitor differences in microbiota composition between samples and between study cohorts. A range of software options are now available, such as mothur, QIIME, VAMPS and GUSTA ME (Schloss et al. 2009; Caporaso et al. 2010; Huse et al. 2014; Buttigieg and Ramette 2015) which allow the researcher to carry out all of the stages involved in processing marker gene survey data, from quality control steps to statistical comparisons and visualisation of results.

While broad marker-gene surveys using universal markers such as the 16S rRNA gene are the most commonly applied variation of this technique it is also possible to carry out focussed surveys of functional genes that have more limited dissemination throughout the microbiota (Walker et al. 2014). The principle here is similar; degenerate PCR primers are targeted towards conserved regions of these functional genes, creating a mixed pool of amplicons, which are then sequenced. This approach has been used, for example, to identify novel groups of butyrate/propionate producing bacteria from the human colon, and cellulolytic bacteria from the rumen (Louis et al. 2010; Reichardt et al. 2014; Brulc et al. 2011). A disadvantage of this targeted approach is that the PCR primers may not efficiently amplify all of the functional genes of interest in a given sample. Untargeted approaches such as metagenomics (see "*Metagenomics*" section below) may circumvent this issue, but at the cost of having to generate far greater amounts of data, which is considerably more expensive to produce and more difficult to analyse (Prakash and Taylor 2012).

### 2.3.2 Whole Genome Sequencing

The first bacterial genome to be completely sequenced was that of *Haemophilus influenzae*, in 1995 (Fleischmann et al. 1995). Then, sequencing was carried out using the traditional Sanger method (Sanger et al. 1992) and it took many years, and hundreds of thousands of dollars, to complete a whole bacterial genome. Advances in DNA sequencing technology since then mean that draft bacterial genomes can now be generated in a matter of hours, and at a cost that is thousands of times cheaper (Loman et al. 2012; Koser et al. 2012). Given the extremely high-throughput nature of next generation sequencing platforms such as Illumina it is now common to simultaneously sequence many microbial genomes on a single sequencing run. This is done by multiplexing samples via the addition of a unique sequence "tags" and then bioinformatically separating reads from each of the combined samples post-sequencing (Lennon et al. 2010). "Shotgun" sequencing, whereby DNA is randomly fragmented prior to sequencing and then the resulting overlapping sequence data is pieced together bioinformatically into contiguous stretches (contigs), is the standard method (Fleischmann et al. 1995). Genomes are typically pieced together by either mapping data on to an existing reference genome (if one is available) or by assembling the data *de novo*. There is a wide range of software available for the genome assembly step, with the optimal choice of assembler depending on the sequencing platform used (Loman et al. 2012).

There are now a large, and constantly increasing, number of genomes available from human-associated microbes. The Human Microbiome Project alone, for example, aims to have generated over 3000 draft genomes once the first phase is complete (Human Microbiome Project Consortium 2012b). Genome sequence data provides critical information on the putative functional capabilities of a given species, although it should be acknowledged that there are often a large number of unannotated genes due to paucity of close, well-characterised matches in reference databases. Indeed, even with *E. coli* K-12, which has been extensively studied and used as a model organism over many decades, around a quarter of the constituent genes remain unannotated (Conway et al. 2014). Nonetheless, as reference databases expand, and techniques for high-throughput, genome-wide, functional probing such as transposon insertion sequencing are developed (van Opijnen and Camilli 2013), this situation will improve. A further, and flourishing, use for whole genome sequencing is in the field of epidemiology, and there are now numerous examples of using whole genome sequence data to trace both global and local dissemination of microbes within human populations (Parkhill and Wren 2011; Eppinger et al. 2014), and to monitor evolutionary changes in genomic content (He et al. 2010; Schuenemann et al. 2013).

### 2.3.3 Metagenomics

An important limitation of whole genome sequencing is that it typically requires the organism to be grown in culture first, so that enough DNA can be extracted for subsequent sequencing. 16S rRNA gene-based surveys have revealed, however, that the majority of human microbiota species have yet to be cultivated in the laboratory (Eckburg et al. 2005). As a result complementary methods such as metagenomics, which can provide genomic insights into this uncultured majority, are attractive options, and have gained increasing favour in recent years. With metagenomics, the researcher directly shotgun sequences DNA extracted from an environmental sample. They then either attempt to bioinformatically piece together the resulting sequence data, which will be comprised of fragments of DNA derived from the range of different species that were present in the original sample, into contiguous stretches of sequence data derived from each individual constituent species, or use the unassembled sequence data directly as a means to assess the functional capabilities of the microbial community as a whole entity (Handelsman 2004).

Metagenomic sequencing in this manner was first applied to samples from the human gut in 2006 (Gill et al. 2006), and has since been used numerous times to study the human microbiota. This technique can be hugely powerful, and it is possible to generate in depth profiles of the functional potential of a given microbial community, including uncultured constituents. It is important to note, however, that the high complexity of many human-associated microbial habitats, such as the colon, means that very deep sequencing is often required in order to generate sufficient sequence data from a representative cross-section of the microbes that are present. Luckily, the development of next-generation sequencing platforms such as Illumina mean that this is now possible, and large-scale metagenomics studies incorporating many individual samples are now being carried out (Hu et al. 2013). Metagenomics is also the only technique that allows effective, in depth, monitoring of the viral communities (or "viromes") that are present in the human body as there are no marker genes equivalent to SSU rRNA that are universally detected in all viruses and so can be used for sequence-surveys (Minot et al. 2011). Further key advantages of metagenomics over other sequence-based techniques are outlined in Table 2.1.

There are, however, some important limitations to the use of metagenomics. For example, this sort of study is far more expensive than marker gene surveying, and comes with a requirement for appropriate computational infrastructure and expertise in order to be able to process the data effectively. Unfortunately, these factors mean that sample sizes tend to be quite small, and large-scale metagenomics studies are currently out of reach for many laboratories. This situation will likely improve though as sequencing costs fall and the use of cloud computing facilities becomes more wide-spread (Angiuoli et al. 2011). As with other DNA-based approaches, the sample storage, preparation, and processing methodologies used will also have significant impacts on the quality of the final metagenomics data (see section "*Common pitfalls of sequence based approaches*" below).

The task of assembling genomes from such a complex collection of microbes, where there will also be great divergences in genome coverage depth based on the relative abundance of each species in the original sample, is also daunting, particularly when trying to assemble genomes from closely related strains and species, or highly fragmented genomes where there is only limited coverage (Nielsen et al. 2014). Although these issues have still not been completely surmounted, there have been great improvements in this area in recent years, and various bioinformatics tools have been developed to aid the genome assembly and species assignment processes (Peng et al. 2011; Namiki et al. 2012; Bankevich et al. 2012; Alneberg et al. 2014).

A further concern is that the current reference databases that are routinely used to classify the DNA sequences are not comprehensive enough. As a result, a large fraction of metagenomics data often goes uncharacterised as there are simply no close matches in the reference database to base a classification on (Thomas et al. 2012). This also means that results tend to be heavily weighted towards well characterised housekeeping genes, which are comparatively well covered in reference databases (Walker et al. 2014). This situation will improve, however, as novel gene functions and pathways are continually elucidated, and reference databases incorporate genomes from a more phylogenetically diverse array of isolates (Walker 2014).

### 2.3.4 Single-Cell Genomics

Single cell genomics (SCG) is an emerging and complementary technique to metagenomics, and is a more targeted approach to generating genomes from uncultured microbes. With this technique, individual microbial cells are isolated from environmental samples, and their genomic DNA subsequently amplified by a whole genome amplification technique (typically multiple displacement amplification) (Walker and Parkhill 2008). This exquisitely powerful amplification step generates sufficient DNA from just a single cell that subsequent shotgun sequencing becomes feasible (Blainey 2013). Moreover, combining SCG with a form of targeted cell selection, such as fluorescent *in situ* hybridisation (Amann and Fuchs 2008), stable isotope probing or Raman microspectroscopy, allows the researcher to potentially recover specific cells that are derived from a particular phylogenetic background, or that carry out a function of interest. As such, SCG complements metagenomics by allowing recovery of genomic information from species that may be rare in the microbial community, and allows the researcher to understand which organisms are capable of carrying out a particular function, even if the genes that are responsible for carrying out this function are unknown or missing from reference databases (Walker et al. 2014).

There are some important limitations to this technique however (see Table 2.1), which have so far hindered wide-scale implementation. Of particular relevance are the issues of contamination (with such a small starting DNA input, any amount of contaminating DNA can easily overwhelm the sequence data that is derived from the cell of interest), and of biases introduced during the amplification step, which can confound genome assembly software, and typically mean that only partial genome coverage can be achieved (Raghunathan et al. 2005). Nonetheless, SCG has been used to characterise novel human-associated bacteria from rare and understudied phyla such as TM7 and *Chloroflexi* (Marcy et al. 2007; Campbell et al. 2014) and holds great promise for wider future application. Results generated can also greatly aid metagenomics-based analyses by broadening reference databases and providing reference genomes to aid with the assembly steps (Rinke et al. 2013).

### 2.3.5 Metatranscriptomics

A further emerging sequence-based technique with applicability to the human microbiota is metatranscriptomics (also termed RNA-seq). Transcriptomics is the study of the RNA transcripts produced by a given species, whereas metatranscriptomics is the study of combined transcripts from an entire microbial community. Thus, in contrast to metagenomics, metatranscriptomics allows insights into the functional *activity* of the microbiota at a given time and under prevailing environmental conditions, not just the functional *potential*. Typically, this technique involves isolating RNA from environmental samples and using this to create reverse transcribed cDNA libraries, which can then be shotgun sequenced using modern high throughput sequencing platforms such as Illumina (Reck et al. 2015). Shotgun sequenced data is then typically assembled by either mapping back to reference genomes, or by carrying out *de novo* assembly. Recent RNA-seq developments now allow strand-specific identification of transcripts, permitting enhanced detection of both messenger and non-coding RNAs, and providing new insights into the roles that the latter may play in cellular function (Croucher and Thomson 2010).

Metatranscriptomics is considerably more technically challenging than metagenomics as it requires additional processing steps such as creating cDNA and depleting host and bacterial rRNAs, which typically make up the vast majority of RNA present in a sample (Giannoukos et al. 2012). Furthermore, transcriptomics is also commonly used in combination with reference genomes, as mapping transcripts back to a reference allows the researcher to understand how a given species responds to changes in environmental conditions. Metatranscriptomic analyses of human microbiota samples are therefore rendered more complex by the fact that there are often no reference genomes available for many members of the microbial community. As such, the raw data may require complex *de novo* assembly prior to analyses, a process which has been improved in recent years with the advent of novel software programmes (Tjaden 2015).

A key limitation of metatranscriptomics is that, due to the very short half-life of mRNA molecules (typically measured in minutes (Reck et al. 2015)), it may not always be entirely representative of microbial activities *in situ*. For example, microbial transcriptional activities measured in faecal samples may not be reflective of gene expression occurring in areas such as the proximal colon. A further limitation is that, as with metagenomics, many of the transcribed genes will be of unknown function due to extensive gaps in reference databases.

Given these inherent complexities and limitations, metatranscriptomics has yet to be applied to human microbiota samples to the same extent as metagenomics, although uptake of this technique is increasing (Jorth et al. 2014; Leimena et al. 2013; Maurice et al. 2013; Macklaim et al. 2013). Moreover, direct comparisons between metagenomic and metatranscriptomic datasets demonstrate the worth of this approach, as highly significant differences between the two datasets are detected, reflective of the fact that microbes are constantly altering their gene expression profiles in response to prevailing environmental conditions (Franzosa et al. 2014).

## 2.4 Common Pitfalls of Sequence Based Approaches

While sequence-based approaches have undoubtedly revolutionised the field of microbiota research there are a number of key caveats, particularly in the areas of sample handling and processing, that should be considered when applying them. Analyses of mock bacterial communities prepared for the Human Microbiome Project, for example, showed that samples clustered together based upon which of four sequencing centres generated the data, illustrating the impact that sample processing steps can have on final sequencing results (Schloss et al. 2011). Furthermore, it is clear from comparisons between techniques that sequence-based approaches commonly "miss" a significant fraction of species present in a sample due to their inherent biases (Shade et al. 2012; Lagier et al. 2012). Awareness of these inherent limitations and biases is therefore important to ensure that erroneous conclusions are not drawn from sequence data (Degnan and Ochman 2012).

Sample preservation is a critical, and often under looked, first step. Emerging evidence suggests that prior freezing of faecal samples can lead to systematic distortions in molecular profiling results. Specifically, it appears that *Bacteroides*-derived DNA may be gradually depleted if samples have been previously held in long term frozen storage (Maukonen et al. 2012; Bahl et al. 2012). Furthermore, evidence suggests that bacterial community profiles obtained from sputum samples may be perturbed by being kept for greater than 12 h at room temperature prior to being placed in long-term frozen storage, and also by repeated freeze-thaw cycles prior to DNA extraction and sequencing (Cuthbertson et al. 2014, 2015).

DNA extraction is another key step, and it is known that choice of extraction kit/method can have major impacts on the final sequencing results obtained (Ferrand et al. 2014; Kennedy et al. 2014). If the chosen DNA extraction method is not robust enough to break open the cell walls of certain microbes then DNA from these species will not be recovered and so will not be observed in the final sequencing libraries. For this reason kits with only chemical-based extraction are not recommended, as they typically generate results with an over-abundance of the more easily extracted Gram negative organisms present in a sample compared to the more recalcitrant Gram positive organisms, which have a stronger cell wall that is less likely to be broken down by chemical lysis only (Walker et al. 2015). DNA extraction kits with a mechanical lysis, or bead-beating, step, which is far more effective at breaking open Gram positive cell walls, are therefore typically recommended (de Boer et al. 2010). However, it should be noted that some bead-beating kits are more effective than others (see Fig. 2.4a) (Kennedy et al. 2014).

For sequence-based approaches requiring prior amplification of specific genes, such as 16S rRNA genes, PCR primer design is a further critically important consideration. It is known that certain groups, for example the Actinobacteria, are systematically under-represented in studies using the commonly used 27f primer (Frank et al. 2008). An example of this is the *Bifidobacterium* genus, typically the dominant member of the gut microbiota in breast fed infants, which has three mismatches to 27f (Fig. 2.4b), therefore this primer should not be used with infant faecal samples as results will not reflect the true microbiota content (Walker et al. 2015). Incorporating degenerate bases into primer design is one way to effectively widen the range of target organisms (Fig. 2.4b). Sim et al (2012), for example, were able to show that improved primers resulted in far better recovery of bifidobacterial sequences from infant faecal samples (Sim et al. 2012).

Primer choice is also important if there are specific groups of bacteria that a researcher is interested in. Next generation sequencing platforms currently generate relatively short reads, meaning that it is typical to target sub-sections of the 16S rRNA gene. Unfortunately, no specific variable region, or combination of variable regions, is able to fully capture the diversity that can be described with full-length 16S rRNA gene sequences. It is therefore prudent to ensure that the species of interest can be differentiated using the variable regions targeted prior to initiating a study (Fig. 2.4c).

A further complicating factor with amplification-based approaches such as marker gene surveys and single-cell genomics is that chimeric molecules can be created during the amplification step (Edgar et al. 2011). Indeed, it is estimated that a significant proportion of DNA sequences submitted to 16S rRNA gene databases, for example, may in fact be chimeric in nature (Ashelford et al. 2005). Chimeric molecules inflate microbial diversity estimates (Schloss et al. 2011), and in the case of single-cell genomics can confound genome assembly software (Lasken and Stockwell 2007). Errors generated during the sequencing process itself can also vastly inflate diversity measures if steps are not taken to account for their impact (Huse et al. 2010). Repeated PCR cycling may also lead to an over-representation of some groups and the under-representation of others. For this reason it has been recommended that the number of PCR cycles should be kept as low as is feasible (Bonnet et al. 2002).

A further potential pitfall is the presence of contamination. Sequence-based approaches are exquisitely sensitive, which means they are an attractive means with which to investigate areas of the body traditionally thought of as “sterile”, or that have very low abundance of colonising microbes that are difficult to grow. Unfortunately, contaminating DNA or cells can be introduced to the sample of interest at many processing stages, including from reagents in common laboratory DNA extraction and PCR kits (Tanner et al. 1998) (Fig. 2.4d). Recent work by Salter et al has indicated that, when sequencing is applied to low biomass samples (i.e. sample containing less than $10^4$ cells), background contamination effectively “swamps” the targeted DNA from the sample and becomes the dominant feature of sequencing results (Salter et al. 2014). Therefore, any researcher working with low biomass samples should ideally make use of copious “negative” sequencing controls. This involves running “blank” DNA extractions and PCR reactions with no sample or template added, and then sequencing these alongside the samples of interest. Any contaminating species detected in the negative controls can then be removed from the sequencing results from the actual samples.

The choice of DNA sequencing platform is a further important consideration. A recent comparative analysis between the Illumina MiSeq and Ion Torrent platforms, for example, indicated that a peculiarity of the Ion Torrent sequencing process can lead to premature truncation of sequence reads derived from certain microbial groups. The effect of this would be to

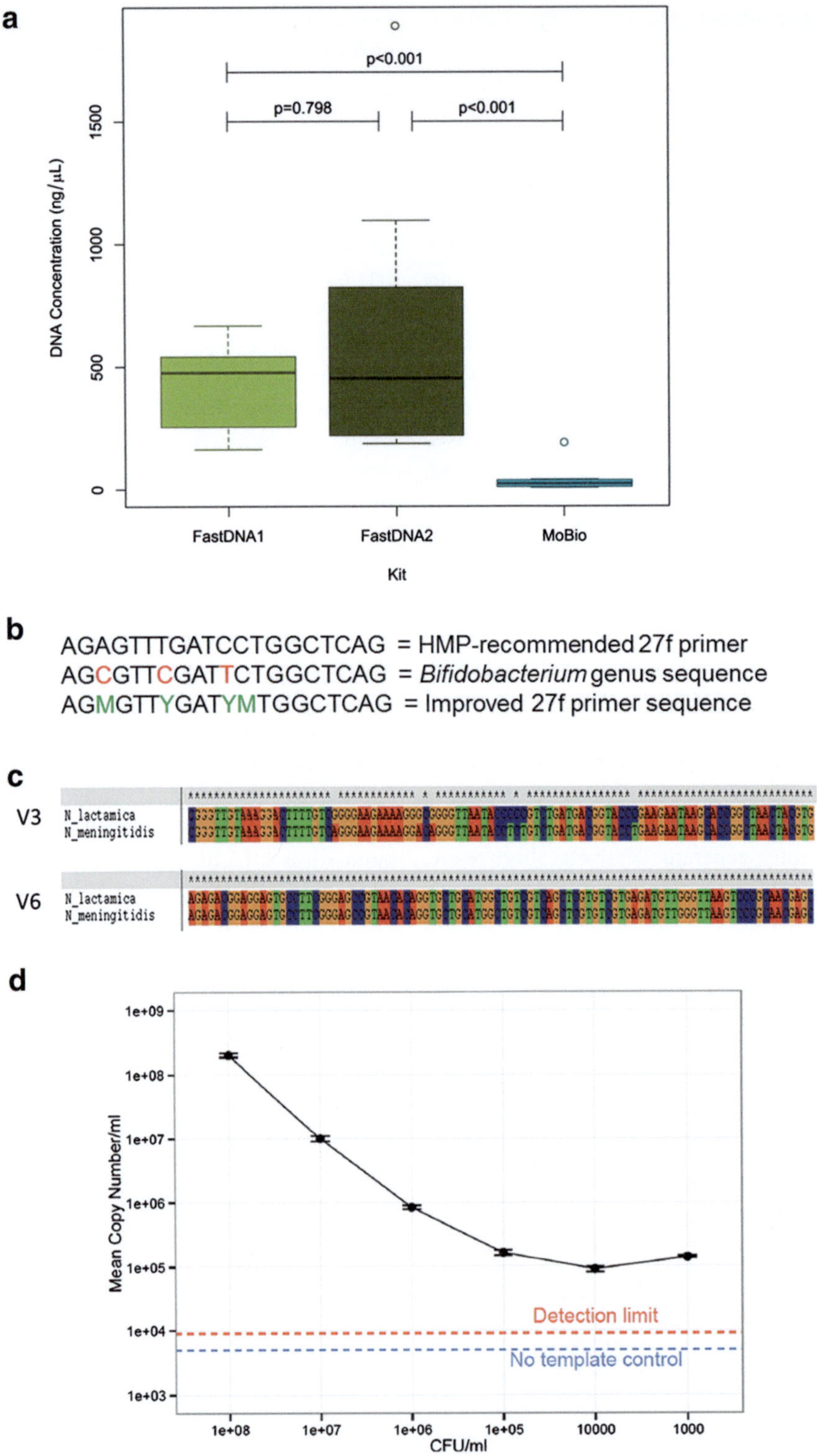


**Fig. 2.4 Importance of optimising sample processing protocols**
**(a)** DNA extraction methodology can impact recovery of DNA from microbiota samples. In this example it can be seen that the yield from a MoBio kit-based protocol is much lower than that from two variations of the FastDNA kit-based protocols. This panel is reprinted in unmodified form from: Kennedy NA et al. The impact of different

bias the results against these groups, and to therefore give misleading estimates of their presence and/or abundance in the original samples. Furthermore, error rates appear to be higher on the Ion Torrent platform, which would artificially inflate measures of diversity (Salipante et al. 2014). The common practise of multiplexing many samples together on a single DNA sequencing run can also introduce bias to the PCR step (Berry et al. 2011) and lead to problems with misidentification of barcoded samples (Esling et al. 2015).

Finally, as DNA can persist in the environment after the death of the host organism, sequencing results (aside from perhaps metatranscriptomics, due to the short half-life of RNA compared to DNA) are unable to distinguish between live and dead/inactive microbes. Results may not therefore accurately represent the active microbiota at the site of interest. However, pre-treatment of samples with agents such as propidium monazide, which can bind to free DNA, and DNA contained within dead or damaged cells, make it possible to make sequencing results more representative of the living or active populations within the microbiota (Rogers et al. 2013).

The combined influence of all of these potentially confounding factors should be particularly borne in mind when conducting meta-analyses incorporating data generated across many different studies where different methodologies have been used since they have the potential to have a greater influence on results obtained than any underlying experimental variable (Wesolowska-Andersen et al. 2014).

## 2.5 Other Community Profiling Approaches

### 2.5.1 Community Fingerprinting Techniques

Due to their falling costs and increased output sequence-based approaches have become the most widely adopted microbial community profiling techniques in recent years. Nonetheless, there are other molecular techniques, such as temperature/denaturing gradient gel electrophoresis (T/DGGE) (Muyzer et al. 1993), terminal restriction fragment length polymorphism (T-RFLP) (Marsh 1999) and automated ribosomal intergenic spacer analysis (ARISA) (Popa et al. 2009), that allow rapid profiling of human-associated microbial communities. These approaches are termed community fingerprinting techniques since they usually give representative overviews of the species present in a sample, without providing direct detailed information about the actual species present. Thus, although these approaches are relatively quick and cheap, the resolution and sensitivity is often much lower than that obtained with direct DNA sequencing (Kovacs et al. 2010; Kisand and Wikner 2003).

**Fig. 2.4** (continued) DNA extraction kits and laboratories upon the assessment of human gut microbiota composition by 16S rRNA genesequencing. (Kennedy et al. 2014) under Creative Commons Attribution (CC BY) license
**(b)** Primer sequence can impact the recovery of species in 16S rRNA gene surveys. In this example it can be seen that the commonly used 27f primer has three mismatches with the important intestinal genus *Bifidobacterium*. As a result this genus is often under-represented in DNA sequence libraries. The bottom configuration shows the same primer with four degenerate bases, which widens the specificity of the primer and improves coverage of groups such as the bifidobacteria (Walker et al. 2015)
**(c)** Choice of 16S rRNA gene variable region can impact species-specificity of sequence results. In this example the V3 region allows differentiation of two *Neisseria* species (*N. meningitidis* and *N. lactamica*) but the sequences from both species are identical over the V6 region, meaning differentiation would not be possible. Therefore, if the researcher was particularly interested in distinguishing these two species, primers targeting the V6 region could not be used
**(d)** Contamination in laboratory reagents. The panel shows the qPCR quantification of a serial dilution of a pure culture of *Salmonella bongori*. The bacterial quantification should reduce in a linear manner as the number of target cells reduces. Instead, the quantification plateaus after three dilutions, indicating the presence of background foreign contamination in the DNA extraction. This panel is reprinted in unmodified form from: Salter et al. (2014) under Creative Commons Attribution (CC BY) license

Although these techniques are gradually falling out of favour, recent work suggests that, while they are not as sensitive as modern next-generation sequencing, they can still generate broadly robust results (van Dorst et al. 2014). It should also be noted that, as they are all DNA extraction and PCR-dependent, and typically make use of marker genes such as 16S rRNA genes, they share many of the limitations and biases of the sequence-based approaches outlined previously in the "*Common pitfalls of sequence based approaches*" section, and in Table 2.1.

### 2.5.2 Microarrays

A microarray is a grid-like collection of microscopic spots of DNA that are anchored to a solid surface. These can be used to probe for the presence of complementary stretches of DNA extracted from a sample of interest by hybridising against the array. Microarrays can therefore be designed to be used in a number of different ways, for example to monitor changes in gene expression, or to mine for the presence of particular functional or marker genes (Paliy and Agans 2012; Tu et al. 2014). Phylogenetic microarrays (sometimes also referred to as phylochips) are a profiling method used in human microbiota research. This technique typically involves creating custom arrays seeded with short oligonucleotides (usually targeting the SSU rRNA genes) that are selected so that they collectively encompass the taxonomic range of organisms expected to be present within a given environmental sample type (Loy et al. 2010). DNA is extracted from the sample of interest, the SSU rRNA genes PCR amplified and labelled with a fluorescent marker and then hybridised against the microarray. When particular DNA spots on the array retain a positive fluorescent signal post-hybridisation, this indicates that the targeted taxonomic group is present in the original sample. By measuring the relative strength of the signal obtained for each positive spot post-hybridisation it may also be possible to semi-quantitatively assess the abundance of different taxa in a sample (Rajilic-Stojanovic et al. 2009).

A potential advantage that the microarray approach has over other profiling techniques is that it typically allows the researcher to simultaneously detect the presence of even quite low abundance organisms, which may not be detected reliably with even a sequence-based approach unless very deep sequencing is carried out. One major limitation though is that, unlike random sequencing approaches, detection is of course limited to the organisms that are targeted by the range of probes that are included on the initial array. Fortunately, there are now comprehensive custom arrays for a range of human-associated habitats such as the gut (Rajilic-Stojanovic et al. 2009; Ladirat et al. 2013; Tottey et al. 2013), vaginal tract (Gautam et al. 2015) and oral cavity (Crielaard et al. 2011), and the range of oligonucleotide probes that are included in these can be expanded as novel species are detected using sequence-based approaches (Rajilic-Stojanovic et al. 2009). It can also be difficult to design arrays where the hybridisation conditions are standardised for all of the probes included. As such it is prudent to control for potential false positives/negatives by including more than one probe for each taxonomic group targeted (Roh et al. 2010). Microarrays also share the same methodological limitations associated with the DNA extraction and PCR steps as other DNA-based techniques (see "*Common pitfalls of sequence based approaches*" section, and Table 2.1).

## 2.6 Quantitative Approaches

There are two widely used molecular methods, namely quantitative PCR and fluorescent *in situ* hybridisation, that allow the enumeration or quantification of dominant groups of microbes within the microbiota. For both of these techniques 16S rRNA gene sequences are typically the underlying basis, with different variable regions targeted with oligonucleotide probes and primers that are specific for particular phylogenetic groups.

### 2.6.1 Quantitative PCR

Quantitative PCR (qPCR), sometimes also referred to as real-time PCR, is a technique based on measuring fluorescence released during PCR amplification (Malinen et al. 2003). The amount of fluorescent signal generated, and the rate at which it accumulates, as the number of PCR cycles increases allows the researcher to quantify the amount of targeted DNA present in a given extraction. This approach is often used to quantify total bacterial cell numbers in a sample, but it can also be used to concurrently quantify the population levels of a number of different bacterial groups by using a range of targeted primer sets (Ramirez-Farias et al. 2009). This is a highly sensitive method and cell densities as low as $10^1$ to $10^3$ cells per sample may be accurately detected (Ott et al. 2004). One limitation of qPCR, however, is that it only allows monitoring of groups that have been specifically targeted by the chosen PCR primers. As a result, untargeted groups will not be observed in the results, and extensive monitoring of microbial communities typically requires the use of multiple different primer sets. Recent efforts have therefore been made to make this approach more high-throughput (Hermann-Bank et al. 2013). Primers must also be extensively tested first, to rule out non-specific binding to non-target DNA. As with all other DNA-based approaches, qPCR is also highly dependent on the choice of DNA extraction methodology.

### 2.6.2 Fluorescent *in situ* Hybridisation (FISH)

FISH is another widely used quantitative technique, with the added advantage that it does not require a DNA extraction step so is free from some of the biases associated with DNA-based methodologies. With FISH, bacterial cells are first fixed using chemicals such as paraformaldehyde and then permeabilised to allow access of fluorescently-labelled oligonucleotide probes. These oligonucleotides are typically between 15 and 30 bases in length and are commonly designed to target regions of rRNA that are specific for chosen phylogenetic groups of bacteria (Amann and Fuchs 2008). Probes may be targeted towards a broad range of bacteria by selecting a highly conserved section of the 16S rRNA gene or towards a narrower range by targeting more specific stretches of the gene (Amann and Ludwig 2000). After entering the fixed cell, the probes hybridise to any sequence of rRNA that is complementary to that of their own. As ribosomes are highly abundant, and distributed throughout the bacterial cell, the targeted cell fluoresces, which allows direct visualisation and enumeration by epifluorescent microscopy (Harmsen et al. 2002). FISH therefore, as well as being a quantitative approach, has the singular advantage that it allows observation of cells of interest *in situ*. For example, it is possible to determine the composition of specific consortia of microbes present on mucosal surfaces, or on the surfaces of particles (Fig. 2.5). A further strength of this approach is that it can be used to link phylogeny to function by employing it in conjunction with techniques such as microautoradiography (MAR-FISH) (Nielsen et al. 2010), Raman microspectroscopy (Raman-FISH) (Wagner 2009) or Secondary Ion Mass Spectrometry (FISH-SIMS) (Musat et al. 2012).

However, there are some important limitations to the use of FISH. It is a far less sensitive quantitative technique than qPCR because a critical mass of bacterial cells (typically around $10^6$ cells/ml of sample) is required per microscopic field of view for accurate visual enumeration. As a result, FISH is most often used to monitor bacterial populations at broader taxonomic levels as individual species only rarely reach the required density for accurate monitoring (Harmsen et al. 2002). As with qPCR, it should also be noted that a further limitation is that FISH only allows monitoring of the microbial groups specifically targeted with oligonucleotide probes, and results can be confounded by false positive/negative results. It is therefore imperative that all newly designed oligonucleotides be tested for specificity prior to use with samples.

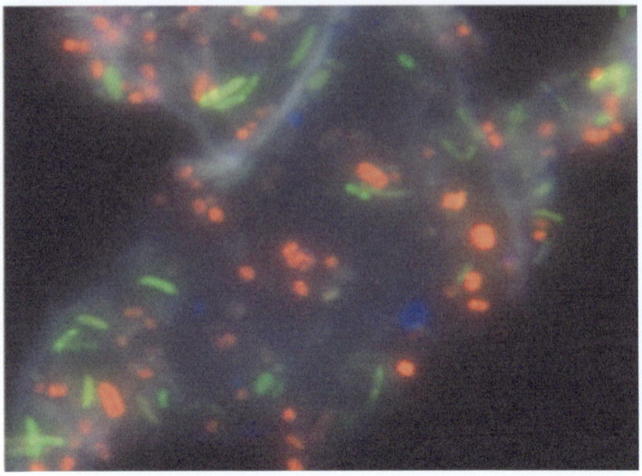

**Fig. 2.5 Fluorescent *in situ* hybridisation** A key advantage of FISH is that it allows direct visualisation of bacteria in environmental samples. In this example we can see groups of bacteria colonising an insoluble fibre particle recovered from a human faecal sample. Cells coloured *green* belong to the *Lachnospiraceae* family, those labelled *red* belong to the *Ruminococcus* genus and those in *blue* are labelled with the universal DAPI stain and do not belong to either of the these bacterial groups. Thus it can be seen that the majority of cells attaching to this fibre are derived from the *Lachnospiraceae* and *Ruminococcaceae*

## 2.7 Functional Analyses

Community profiling techniques can only provide an overview of the microbial composition in a given sample or, in the case of shotgun metagenomics, can only provide an overview of the encoding potential of a microbial ecosystem. Indeed, while we now have a much clearer picture of the kind of microbes that inhabit the various niches associated with the human body we know comparatively far less about the roles that each individual species plays. Fortunately, there are now a number of complementary techniques, beyond the culture-based and metatranscriptomics methods described previously in this chapter, that can be used to assess the functionality of the microbiota.

### 2.7.1 Functional Metagenomics

In contrast to whole shotgun metagenomics, where the aim is to generate deep sequencing-based profiles of the entire functional capability of the microbiota, with functional metagenomics the aim is instead to identify specific functional genes by cloning and expressing them in a surrogate bacterial species (Handelsman et al. 1998). Typically this involves large-scale cloning of random environmental DNA fragments into a host species such as *E. coli* and then screening for activity by growing the transformed host species on agar plates containing a substrate of interest. Where functional activity is observed, the cloned gene can then be sequenced to provide supporting genomic data. This approach has been used, for example, to identify complex-carbohydrate degrading enzymes derived from the human gut (Tasse et al. 2010). Functional metagenomics is therefore a potentially hugely powerful approach, with the key advantage that it allows the researcher to simultaneously identify novel genes encoding specific functions from a broad range of bacterial species, including those that may not be amenable to culture in the laboratory (Uchiyama and Miyazaki 2009). A further advantage is that the functional annotation of previously unknown genes enhances reference databases, which can then be used to improve classification success rates and accuracy for sequence-based shotgun metagenomics studies.

There are, however, a number of important limitations, which has so far limited the use of functional metagenomics in comparison to the sequence-based shotgun metagenomics approach. For example, it is typically highly laborious and inefficient; millions of random DNA fragments may need to be cloned in order to identify activities of interest. Furthermore, there are important technological barriers that impinge upon the effectiveness of the approach. Many of the cloned fragments will be poorly expressed by foreign hosts such as *E. coli*, meaning that alternative hosts/approaches may need to be considered (Liebl et al. 2014). In addition, the DNA extraction step is crucially important as the researcher must reach a balance between using a protocol that is stringent enough to extract DNA from as wide a range of species in the original sample as possible, but is not so stringent that it shears the resulting DNA to the extent that many of the cloned genes and gene clusters are disrupted (Kakirde et al. 2010). A further limitation is that, while this approach may identify products formed from individual genes or relatively simple contiguous gene clusters, it is unlikely to be able to identify gene products that result from complex metabolic pathways (Walker et al. 2014).

### 2.7.2 Metaproteomics

Metaproteomics is the study of the complement of proteins produced by mixed microbial communities (Wilmes and Bond 2009). As such, it provides functional information by allowing the researcher to monitor changes in protein expression by the entire microbiota in response to changes in prevailing environmental conditions. With this technique, proteins must first be extracted from the environmental sample of interest, and then separated prior to characterisation with mass spectrometry and subsequent bioinformatics-based comparisons with reference databases (Hettich et al. 2012). Until recently proteins (or peptides) were most commonly separated by using gel electrophoresis approaches (Magdeldin et al. 2014), but they are now increasingly separated by using liquid chromatography instead. Recent technological advances in the field mean that it is now possible to carry out very high throughput liquid chromatography-mass spectrometry based analyses, where many thousands of different proteins/peptides can be separated and characterised (Hettich et al. 2013).

Metaproteomics offers some key advantages over the metatranscriptomics approach described previously in that, by measuring proteins rather than mRNA, it provides a broader, more representative picture of the functional activity of the microbiota as it also accounts for the impact of processes such as post-translational modifications (Cain et al. 2014). Proteins are also typically more stable than mRNA molecules, meaning that results obtained may not be so dependent on the speed with which the samples are processed. A particular advantage over DNA-based metagenomics is that metaproteomics is faster, and cheaper (Verberkmoes et al. 2009). The relatively untargeted nature of metaproteomics also means that it may be possible to identify marker proteins that are indicative of a healthy or diseased human host status.

However, although the technology involved in metaproteomics is rapidly improving, there are a range of important limitations, and this technique is currently far less commonly applied in comparison to DNA-based approaches. Although resolution is improving, metaproteomics can only currently characterise thousands out of the millions of proteins/peptides that might be present in a complex microbiota sample at one time (Kolmeder and de Vos 2014). As such, only proteins produced by the most dominant members of the microbiota can be expected to be captured with reasonable coverage (Verberkmoes et al. 2009). It can also be difficult to differentiate similar proteins or ascribe them to particular phylogenetic groups (Lichtman et al. 2015), and, as with metagenomics studies, a large proportion of the data recovered will have no close matches to available reference databases (Verberkmoes et al. 2009). The methodology chosen during the protein extraction step will also have significant impacts on the representativeness of the protein complement recovered, and it is important to extract proteins with reasonable efficiency from both Gram positive and Gram negative constituents (Tanca et al. 2014). Human-derived proteins will also be present, and can be a highly significant component in samples such as biopsies, meaning it is sometimes necessary to carry out selective steps to enrich for microbial proteins (Kolmeder and de Vos 2014). There are also issues surrounding reproducibility between samples, particularly when using gel electrophoresis to separate proteins (Magdeldin et al. 2014).

### 2.7.3 Metabolomics

Metabolomics is the study of the metabolites/small molecules present within a given sample at the time of sampling. As with the metaproteomics approach outlined above, metabolomics therefore offers distinct advantages over other functional approaches such as metatranscriptomics as it allows the direct monitoring of the end products of bacterial metabolism (Ursell et al. 2014). With metabolomics, metabolites are typically isolated from bodily samples such as urine, faeces and blood and measured using technologies such nuclear magnetic resonance (NMR) microscopy or mass spectrometry (Nicholson and Lindon 2008). The end result of these approaches are a series of characteristic spectra or peaks derived from the range of metabolites that are present within the original sample (Savorani et al. 2013). Depending on the approach used, metabolomic screens can either be carried out in a targeted way for particular groups of metabolites (for example, short chain fatty acids), or on a more global basis (Griffiths et al. 2010). In the latter case, the main challenge is to assign particular spectra from the complex mixture of peaks to specific compounds, and then to attempt to correlate presence/absence of these compounds with markers of host health (Lenz and Wilson 2007). By simultaneously capturing both host and microbial-derived metabolites, metabolomics has particular appeal as an approach to characterise host-microbe interactions (Wikoff et al. 2009).

A key limitation of this technique is that it can be difficult to accurately determine which microbial species are producing particular metabolites. While attempts are often made to correlate metabolite production with microbial composition data generated in tandem by sequence survey or metagenomic approaches, these can be confounded by the presence of DNA derived from dead or inactive species in the sequence-based results, and by the fact that there can be considerable metabolic flux within complex ecosystems, such that metabolites associated with taxa that are dominant in sequence surveys may not actually be produced by them (Abram 2015). Furthermore, many metabolites, for example short chain fatty acids, are rapidly absorbed by the host, meaning that production levels cannot be accurately defined or ascribed to particular species (Kolmeder and de Vos 2014). An additional important disadvantage is that reference databases are generally lacking, even more so than those for DNA and proteins, meaning that only a small fraction of metabolomics data can currently be assigned to known metabolites (Baker 2011). Finally, as with metaproteomics, resolution limits (even with the most modern instruments) mean that it is only possible to accurately monitor a small subset of the wide range of metabolites that may be present in a complex sample such as faeces (Goedert et al. 2014).

It can be seen, therefore, that all four key modern "omics" technologies (metagenomics for DNA, metatranscriptomics for RNA, metaproteomics for proteins, and metabolomics for metabolites) have distinct strengths and limitations. As a result, there is increasing interest in integrating the output from each of these approaches in order to enhance their overall power and provide a more comprehensive, systems biology-based, overview of the human microbiota. Effective integration of these complex datasets remains to some extent an unfulfilled ambition, but one that is being rapidly guided by improvements in computing infrastructure, bioinformatics, mathematical modelling and statistical approaches (Abram 2015).

### 2.7.4 Stable Isotope Probing

One final functional approach with strong applicability to the study of the human microbiota is stable isotope probing (SIP). With this technique, mixed microbial communities are incubated with labelled substrates containing heavy stable isotopes such as $^{13}C$, $^{15}N$, and $^{18}O$. Species that are able to grow on the labelled substrate incorporate the isotope markers into cellular biomass, which can then be studied by looking at components such as DNA (DNA-SIP), RNA (RNA-SIP), proteins (protein-SIP) or phospholipid-derived fatty acids (PFLA-SIP). Approaches like density gradient ultracentrifugation (Dunford and Neufeld 2010) or advanced single-cell resolution techniques such as Raman microspectroscopy and Secondary Ion Mass Spectrometry (SIMS) are used to distinguish the active microbes from species that did not incorporate the marker (Eichorst et al. 2015). Regardless of the actual cellular components targeted SIP is therefore an attractive basis for uncovering which microbes within complex microbial communities carry out particular functions (Uhlik et al. 2013).

SIP is an emerging means with which to unravel the complex activities of the human microbiota. Early studies used this technique in tandem with community profiling approaches like T-RFLP and FISH to characterise the microbes that were able to actively utilise labelled substrates such as resistant starch and oligofructose (Kovatcheva-Datchary et al. 2009; Reichardt et al. 2011). When used in combination with more modern "omics" techniques SIP has the potential to be particularly powerful. For example, fractionated DNA or RNA containing the stable isotopes can then be sequenced using marker gene surveys, metagenomics or metatranscriptomics in order to identify the species that were active during incubation with the labelled substrate (Chen and Murrell 2010). Similarly, advances in micro-manipulation technologies such as optical tweezers mean that whole cells that have been shown by techniques like Raman microspectroscopy to have incorporated the stable isotopes can then be isolated from the sample and either cultured or, if that is not possible, put forward for genome sequencing via single cell genomics (Berry et al. 2015).

While SIP approaches can be hugely powerful there are important caveats, which have limited widespread application of these techniques thus far. SIP is far more technically challenging than approaches such as SSU rRNA gene surveys or metagenomics, and the modern single-cell resolution techniques such as Raman microspectroscopy and SIMS can be prohibitively expensive (Wagner 2009). Similarly, use of SIP is limited by the supply and cost of labelled substrates (Uhlik et al. 2013). Recent innovations though, such as the use of the cheap and readily available heavy water ($D_2O$) as a general marker of cellular growth, allows SIP to be carried out without specific labelled carbon or nitrogen sources (Berry et al. 2015). A further limitation is that SIP requires microbes to be grown in the presence of the labelled tracer so that it can be incorporated into active cells. Often this means growing mixed communities under artificial laboratory conditions, meaning that results may not entirely reflect the activity of the microbiota *in vivo* (Uhlik et al. 2013). Nonetheless, impressive new innovations have, for example, allowed researchers to identify microbes growing *in vivo* that forage host-derived proteins for growth (Berry et al. 2013). Finally, there is considerable metabolic flux within complex microbial communities, with cross feeding between species a common feature. This means that stable isotopes such as $^{13}C$ may "flow" from the primary degrader of a labelled substrate to many other species that are present within the community, potentially impeding the ability to detect the initial utilising species (Dumont and Murrell 2005).

## 2.8 Conclusions

There are now many different ways in which the human microbiota can be studied, and each methodology has inherent advantages and limitations. Ultimately, the best technical approach for a given situation will clearly depend on the question that the researcher wishes to address. Although each technique has largely been considered in isolation in this review it should be emphasised here that, where possible, the synergistic use of multiple methodological approaches offers perhaps the greatest power with which to uncover novel insights.

Looking towards the future, it is clear that further improvements in sequence-based technologies, molecular methods, model systems and bioinformatics will continue to open up novel avenues for research. The synergistic adoption of such approaches will greatly enhance our ability to take a systems biology-based view of the human microbiota, and how it interacts with the host. Traditional techniques such as culture will also retain an important role as we seek to translate omics-based observations into interventions such as probiotics and pharmabiotics aimed at improving host health (Reardon 2014). We have come a long way since Antonie van Leeuwenhoek's first glimpses of the human microbiota, and are now quickly entering an era where our increased understanding of our microbial inhabitants is being put to practical therapeutic use (Shanahan 2015). Further technological advances can only accelerate this process.

**Acknowledgements** The author, and The Rowett Institute of Nutrition and Health, University of Aberdeen, receives core funding support from the Scottish Government Rural and Environmental Science and Analysis Service (RESAS).

## References

Abram F (2015) Systems-based approaches to unravel multi-species microbial community functioning. Comput Struct Biotechnol J 13:24–32

Alneberg J, Bjarnason BS, de Bruijn I, Schirmer M, Quick J, Ijaz UZ et al (2014) Binning metagenomic contigs by coverage and composition. Nat Methods 11(11):1144–1146

Amann R, Fuchs BM (2008) Single-cell identification in microbial communities by improved fluorescence *in situ* hybridization techniques. Nat Rev Microbiol 6(5):339–348

Amann R, Ludwig W (2000) Ribosomal RNA-targeted nucleic acid probes for studies in microbial ecology. FEMS Microbiol Rev 24(5):555–565

Angiuoli SV, White JR, Matalka M, White O, Fricke WF (2011) Resources and costs for microbial sequence analysis evaluated using virtual machines and cloud computing. PLoS One 6(10), e26624

Ashelford KE, Chuzhanova NA, Fry JC, Jones AJ, Weightman AJ (2005) At least 1 in 20 16S rRNA sequence records currently held in public repositories is estimated to contain substantial anomalies. Appl Environ Microbiol 71(12):7724–7736

Bahl MI, Bergstrom A, Licht TR (2012) Freezing fecal samples prior to DNA extraction affects the Firmicutes to Bacteroidetes ratio determined by downstream quantitative PCR analysis. FEMS Microbiol Lett 329(2):193–197

Baker M (2011) Metabolomics: from small molecules to big ideas. Nat Methods 8:117–121

Bankevich A, Nurk S, Antipov D, Gurevich AA, Dvorkin M, Kulikov AS et al (2012) SPAdes: a new genome assembly algorithm and its applications to single-cell sequencing. J Comput Biol 19(5):455–477

Belenguer A, Duncan SH, Calder AG, Holtrop G, Louis P, Lobley GE et al (2006) Two routes of metabolic cross-feeding between Bifidobacterium adolescentis and butyrate-producing anaerobes from the human gut. Appl Environ Microbiol 72(5):3593–3599

Berry D, Ben Mahfoudh K, Wagner M, Loy A (2011) Barcoded primers used in multiplex amplicon pyrosequencing bias amplification. Appl Environ Microbiol 77(21):7846–7849

Berry D, Stecher B, Schintlmeister A, Reichert J, Brugiroux S, Wild B et al (2013) Host-compound foraging by intestinal microbiota revealed by single-cell stable isotope probing. Proc Natl Acad Sci U S A 110(12):4720–4725

Berry D, Mader E, Lee TK, Woebken D, Wang Y, Zhu D et al (2015) Tracking heavy water (D2O) incorporation for identifying and sorting active microbial cells. Proc Natl Acad Sci U S A 112(2):E194–E203

Blainey PC (2013) The future is now: single-cell genomics of bacteria and archaea. FEMS Microbiol Rev 37(3):407–427

Bomar L, Maltz M, Colston S, Graf J (2011) Directed culturing of microorganisms using metatranscriptomics. mBio 2(2):e00012-11

Bonnet R, Suau A, Dore J, Gibson GR, Collins MD (2002) Differences in rDNA libraries of faecal bacteria derived from 10- and 25-cycle PCRs. Int J Syst Evol Microbiol 52(Pt 3):757–763

Brulc JM, Yeoman CJ, Wilson MK, Berg Miller ME, Jeraldo P, Jindou S et al (2011) Cellulosomics, a gene-centric approach to investigating the intraspecific diversity and adaptation of Ruminococcus flavefaciens within the rumen. PLoS One 6(10), e25329

Buttigieg PL, Ramette A (2015) A guide to statistical analysis in microbial ecology: a community-focused, living review of multivariate data analyses. FEMS Microbiol Ecol 90:543–550

Cain JA, Solis N, Cordwell SJ (2014) Beyond gene expression: the impact of protein post-translational modifications in bacteria. J Proteome 97:265–286

Campbell AG, Schwientek P, Vishnivetskaya T, Woyke T, Levy S, Beall CJ et al (2014) Diversity and genomic insights into the uncultured Chloroflexi from the human microbiota. Environ Microbiol 16(9):2635–2643

Caporaso JG, Kuczynski J, Stombaugh J, Bittinger K, Bushman FD, Costello EK et al (2010) QIIME allows analysis of high-throughput community sequencing data. Nat Methods 7(5):335–336

Caporaso JG, Lauber CL, Walters WA, Berg-Lyons D, Lozupone CA, Turnbaugh PJ et al (2011) Global patterns of 16S rRNA diversity at a depth of millions of sequences per sample. Proc Natl Acad Sci U S A 108(Suppl 1):4516–4522

Chen Y, Murrell JC (2010) When metagenomics meets stable-isotope probing: progress and perspectives. Trends Microbiol 18(4):157–163

Chun J, Lee JH, Jung Y, Kim M, Kim S, Kim BK et al (2007) EzTaxon: a web-based tool for the identification of prokaryotes based on 16S ribosomal RNA gene sequences. Int J Syst Evol Microbiol 57(Pt 10):2259–2261

Cole JR, Wang Q, Fish JA, Chai B, McGarrell DM, Sun Y et al (2014) Ribosomal database project: data and tools for high throughput rRNA analysis. Nucleic Acids Res 42(Database issue):D633–D642

Conway T, Creecy JP, Maddox SM, Grissom JE, Conkle TL, Shadid TM et al (2014) Unprecedented high-resolution view of bacterial operon architecture revealed by RNA sequencing. mBio 5(4):e01442-14

Crielaard W, Zaura E, Schuller AA, Huse SM, Montijn RC, Keijser BJ (2011) Exploring the oral microbiota of children at various developmental stages of their dentition in the relation to their oral health. BMC Med Genet 4:22

Croucher NJ, Thomson NR (2010) Studying bacterial transcriptomes using RNA-seq. Curr Opin Microbiol 13(5):619–624

Cryan JF, Dinan TG (2012) Mind-altering microorganisms: the impact of the gut microbiota on brain and behaviour. Nat Rev Neurosci 13(10):701–712

Cuthbertson L, Rogers GB, Walker AW, Oliver A, Hafiz T, Hoffman LR et al (2014) Time between collection and storage significantly influences bacterial sequence composition in sputum samples from cystic fibrosis respiratory infections. J Clin Microbiol 52(8):3011–3016

Cuthbertson L, Rogers GB, Walker AW, Oliver A, Hoffman LR, Carroll MP et al (2015) Implications of multiple freeze-thawing on respiratory samples for culture-independent analyses. J Cyst Fibros 14(4):464–467

de Boer R, Peters R, Gierveld S, Schuurman T, Kooistra-Smid M, Savelkoul P (2010) Improved detection of microbial DNA after bead-beating before DNA isolation. J Microbiol Methods 80(2):209–211

Degnan PH, Ochman H (2012) Illumina-based analysis of microbial community diversity. ISME J 6(1):183–194

DeSantis TZ, Hugenholtz P, Larsen N, Rojas M, Brodie EL, Keller K et al (2006) Greengenes, a chimera-checked 16S rRNA gene database and workbench compatible with ARB. Appl Environ Microbiol 72(7):5069–5072

Dumont MG, Murrell JC (2005) Stable isotope probing – linking microbial identity to function. Nat Rev Microbiol 3(6):499–504

Duncan SH, Hold GL, Harmsen HJ, Stewart CS, Flint HJ (2002) Growth requirements and fermentation products of Fusobacterium prausnitzii, and a proposal to reclassify it as Faecalibacterium prausnitzii gen. nov., comb. nov. Int J Syst Evol Microbiol 52(Pt 6):2141–2146

Dunford EA, Neufeld JD (2010) DNA stable-isotope probing (DNA-SIP). J Vis Exp JoVE 42

Eckburg PB, Bik EM, Bernstein CN, Purdom E, Dethlefsen L, Sargent M et al (2005) Diversity of the human intestinal microbial flora. Science 308(5728):1635–1638

Edgar RC, Haas BJ, Clemente JC, Quince C, Knight R (2011) UCHIME improves sensitivity and speed of chimera detection. Bioinformatics 27(16):2194–2200

Eichorst SA, Strasser F, Woyke T, Schintlmeister A, Wagner M, Woebken D (2015) Advancements in the application of NanoSIMS and Raman microspectroscopy to investigate the activity of microbial cells in soils. FEMS Microbiol Ecol 91(10):fiv106

Ellekilde M, Selfjord E, Larsen CS, Jakesevic M, Rune I, Tranberg B et al (2014) Transfer of gut microbiota from lean and obese mice to antibiotic-treated mice. Sci rep 4:5922

Eller C, Crabill MR, Bryant MP (1971) Anaerobic roll tube media for nonselective enumeration and isolation of bacteria in human feces. Appl Microbiol 22(4):522–529

Eppinger M, Pearson T, Koenig SS, Pearson O, Hicks N, Agrawal S et al (2014) Genomic epidemiology of the Haitian cholera outbreak: a single introduction followed by rapid, extensive, and continued spread characterized the onset of the epidemic. mBio 5(6):e01721

Ericsson AC, Davis JW, Spollen W, Bivens N, Givan S, Hagan CE et al (2015) Effects of vendor and genetic background on the composition of the fecal microbiota of inbred mice. PLoS One 10(2), e0116704

Esling P, Lejzerowicz F, Pawlowski J (2015) Accurate multiplexing and filtering for high-throughput amplicon-sequencing. Nucleic Acids Res 43(5):2513–2524

Ferenci T (1999) 'Growth of bacterial cultures' 50 years on: towards an uncertainty principle instead of constants in bacterial growth kinetics. Res Microbiol 150(7):431–438

Ferrand J, Patron K, Legrand-Frossi C, Frippiat JP, Merlin C, Alauzet C et al (2014) Comparison of seven methods for extraction of bacterial DNA from fecal and cecal samples of mice. J Microbiol Methods 105:180–185

Fleischmann RD, Adams MD, White O, Clayton RA, Kirkness EF, Kerlavage AR et al (1995) Whole-genome random sequencing and assembly of Haemophilus influenzae Rd. Science 269(5223):496–512

Flint HJ, Duncan SH, Scott KP, Louis P (2007) Interactions and competition within the microbial community of

the human colon: links between diet and health. Environ Microbiol 9(5):1101–1111
Frank JA, Reich CI, Sharma S, Weisbaum JS, Wilson BA, Olsen GJ (2008) Critical evaluation of two primers commonly used for amplification of bacterial 16S rRNA genes. Appl Environ Microbiol 74(8):2461–2470
Franzosa EA, Morgan XC, Segata N, Waldron L, Reyes J, Earl AM et al (2014) Relating the metatranscriptome and metagenome of the human gut. Proc Natl Acad Sci U S A 111(22):E2329–E2338
Gautam R, Borgdorff H, Jespers V, Francis SC, Verhelst R, Mwaura M et al (2015) Correlates of the molecular vaginal microbiota composition of African women. BMC Infect Dis 15:86
Giannoukos G, Ciulla DM, Huang K, Haas BJ, Izard J, Levin JZ et al (2012) Efficient and robust RNA-seq process for cultured bacteria and complex community transcriptomes. Genome Biol 13(3):R23
Gill SR, Pop M, Deboy RT, Eckburg PB, Turnbaugh PJ, Samuel BS et al (2006) Metagenomic analysis of the human distal gut microbiome. Science 312(5778):1355–1359
Goedert JJ, Sampson JN, Moore SC, Xiao Q, Xiong X, Hayes RB et al (2014) Fecal metabolomics: assay performance and association with colorectal cancer. Carcinogenesis 35(9):2089–2096
Goodman AL, Kallstrom G, Faith JJ, Reyes A, Moore A, Dantas G et al (2011) Extensive personal human gut microbiota culture collections characterized and manipulated in gnotobiotic mice. Proc Natl Acad Sci U S A 108(15):6252–6257
Griffiths WJ, Koal T, Wang Y, Kohl M, Enot DP, Deigner HP (2010) Targeted metabolomics for biomarker discovery. Angew Chem 49(32):5426–5445
Handelsman J (2004) Metagenomics: application of genomics to uncultured microorganisms. Microbiol Mol Biol Rev 68(4):669–685
Handelsman J, Rondon MR, Brady SF, Clardy J, Goodman RM (1998) Molecular biological access to the chemistry of unknown soil microbes: a new frontier for natural products. Chem Biol 5(10):R245–R249
Harmsen HJ, Raangs GC, He T, Degener JE, Welling GW (2002) Extensive set of 16S rRNA-based probes for detection of bacteria in human feces. Appl Environ Microbiol 68(6):2982–2990
He M, Sebaihia M, Lawley TD, Stabler RA, Dawson LF, Martin MJ et al (2010) Evolutionary dynamics of Clostridium difficile over short and long time scales. Proc Natl Acad Sci U S A 107(16):7527–7532
Hermann-Bank ML, Skovgaard K, Stockmarr A, Larsen N, Molbak L (2013) The Gut Microbiotassay: a high-throughput qPCR approach combinable with next generation sequencing to study gut microbial diversity. BMC Genomics 14:788
Hettich RL, Sharma R, Chourey K, Giannone RJ (2012) Microbial metaproteomics: identifying the repertoire of proteins that microorganisms use to compete and cooperate in complex environmental communities. Curr Opin Microbiol 15(3):373–380
Hettich RL, Pan C, Chourey K, Giannone RJ (2013) Metaproteomics: harnessing the power of high performance mass spectrometry to identify the suite of proteins that control metabolic activities in microbial communities. Anal Chem 85(9):4203–4214
Hu Y, Yang X, Qin J, Lu N, Cheng G, Wu N et al (2013) Metagenome-wide analysis of antibiotic resistance genes in a large cohort of human gut microbiota. Nat Commun 4:2151
Human Microbiome Project C (2012a) Structure, function and diversity of the healthy human microbiome. Nature 486(7402):207–214
Human Microbiome Project Consortium (2012b) A framework for human microbiome research. Nature 486(7402):215–221
Huse SM, Welch DM, Morrison HG, Sogin ML (2010) Ironing out the wrinkles in the rare biosphere through improved OTU clustering. Environ Microbiol 12(7):1889–1898
Huse SM, Mark Welch DB, Voorhis A, Shipunova A, Morrison HG, Eren AM et al (2014) VAMPS: a website for visualization and analysis of microbial population structures. BMC Bioinforma 15:41
Jorth P, Turner KH, Gumus P, Nizam N, Buduneli N, Whiteley M (2014) Metatranscriptomics of the human oral microbiome during health and disease. M Bio 5(2):e01012–e01014
Kakirde KS, Parsley LC, Liles MR (2010) Size does matter: application-driven approaches for soil Metagenomics. Soil Biol Biochem 42(11):1911–1923
Kennedy NA, Walker AW, Berry SH, Duncan SH, Farquarson FM, Louis P et al (2014) The impact of different DNA extraction kits and laboratories upon the assessment of human gut microbiota composition by 16S rRNA gene sequencing. PLoS One 9(2), e88982
Kisand V, Wikner J (2003) Limited resolution of 16S rDNA DGGE caused by melting properties and closely related DNA sequences. J Microbiol Methods 54(2):183–191
Klaasen HL, Koopman JP, Van den Brink ME, Van Wezel HP, Beynen AC (1991) Mono-association of mice with non-cultivable, intestinal, segmented, filamentous bacteria. Arch Microbiol 156(2):148–151
Kolmeder CA, de Vos WM (2014) Metaproteomics of our microbiome – developing insight in function and activity in man and model systems. J Proteome 97:3–16
Koser CU, Ellington MJ, Cartwright EJ, Gillespie SH, Brown NM, Farrington M et al (2012) Routine use of microbial whole genome sequencing in diagnostic and public health microbiology. PLoS Pathog 8(8), e1002824
Kostic AD, Howitt MR, Garrett WS (2013) Exploring host-microbiota interactions in animal models and humans. Genes Dev 27(7):701–718

Kovacs A, Yacoby K, Gophna U (2010) A systematic assessment of automated ribosomal intergenic spacer analysis (ARISA) as a tool for estimating bacterial richness. Res Microbiol 161(3):192–197

Kovatcheva-Datchary P, Egert M, Maathuis A, Rajilic-Stojanovic M, de Graaf AA, Smidt H et al (2009) Linking phylogenetic identities of bacteria to starch fermentation in an in vitro model of the large intestine by RNA-based stable isotope probing. Environ Microbiol 11(4):914–926

Ladirat SE, Schols HA, Nauta A, Schoterman MH, Keijser BJ, Montijn RC et al (2013) High-throughput analysis of the impact of antibiotics on the human intestinal microbiota composition. J Microbiol Methods 92(3):387–397

Lagier JC, Armougom F, Million M, Hugon P, Pagnier I, Robert C et al (2012) Microbial culturomics: paradigm shift in the human gut microbiome study. Clin Microbiol Infect 18(12):1185–1193

Lagier JC, Edouard S, Pagnier I, Mediannikov O, Drancourt M, Raoult D (2015a) Current and past strategies for bacterial culture in clinical microbiology. Clin Microbiol Rev 28(1):208–236

Lagier JC, Hugon P, Khelaifia S, Fournier PE, La Scola B, Raoult D (2015b) The rebirth of culture in microbiology through the example of culturomics to study human gut microbiota. Clin Microbiol Rev 28(1):237–264

Lasken RS, Stockwell TB (2007) Mechanism of chimera formation during the multiple displacement amplification reaction. BMC Biotechnol 7:19

Lees H, Swann J, Poucher SM, Nicholson JK, Holmes E, Wilson ID et al (2014) Age and microenvironment outweigh genetic influence on the Zucker rat microbiome. PLoS One 9(9), e100916

Leimena MM, Ramiro-Garcia J, Davids M, van den Bogert B, Smidt H, Smid EJ et al (2013) A comprehensive metatranscriptome analysis pipeline and its validation using human small intestine microbiota datasets. BMC Genomics 14:530

Lennon NJ, Lintner RE, Anderson S, Alvarez P, Barry A, Brockman W et al (2010) A scalable, fully automated process for construction of sequence-ready barcoded libraries for 454. Genome Biol 11(2):R15

Lenz EM, Wilson ID (2007) Analytical strategies in metabolomics. J Proteome Res 6(2):443–458

Li J, Jia H, Cai X, Zhong H, Feng Q, Sunagawa S et al (2014) An integrated catalog of reference genes in the human gut microbiome. Nat Biotechnol 32(8):834–841

Lichtman JS, Sonnenburg JL, Elias JE (2015) Monitoring host responses to the gut microbiota. ISME J 9(9):1908–1915

Liebl W, Angelov A, Juergensen J, Chow J, Loeschcke A, Drepper T et al (2014) Alternative hosts for functional (meta) genome analysis. Appl Microbiol Biotechnol 98(19):8099–8109

Loman NJ, Constantinidou C, Chan JZ, Halachev M, Sergeant M, Penn CW et al (2012) High-throughput bacterial genome sequencing: an embarrassment of choice, a world of opportunity. Nat Rev Microbiol 10(9):599–606

Louis P, Young P, Holtrop G, Flint HJ (2010) Diversity of human colonic butyrate-producing bacteria revealed by analysis of the butyryl-CoA:acetate CoA-transferase gene. Environ Microbiol 12(2):304–314

Loy A, Pester M, Steger D (2010) Phylogenetic microarrays for cultivation-independent identification and metabolic characterization of microorganisms in complex samples. Methods Mol Biol 688:187–206

Macklaim JM, Fernandes AD, Di Bella JM, Hammond JA, Reid G, Gloor GB (2013) Comparative meta-RNA-seq of the vaginal microbiota and differential expression by Lactobacillus iners in health and dysbiosis. Microbiome 1(1):12

Magdeldin S, Enany S, Yoshida Y, Xu B, Zhang Y, Zureena Z et al (2014) Basics and recent advances of two dimensional- polyacrylamide gel electrophoresis. Clin proteomics 11(1):16

Malinen E, Kassinen A, Rinttila T, Palva A (2003) Comparison of real-time PCR with SYBR Green I or 5′-nuclease assays and dot-blot hybridization with rDNA-targeted oligonucleotide probes in quantification of selected faecal bacteria. Microbiology 149(Pt 1):269–277

Marcy Y, Ouverney C, Bik EM, Losekann T, Ivanova N, Martin HG et al (2007) Dissecting biological "dark matter" with single-cell genetic analysis of rare and uncultivated TM7 microbes from the human mouth. Proc Natl Acad Sci U S A 104(29):11889–11894

Marsh TL (1999) Terminal restriction fragment length polymorphism (T-RFLP): an emerging method for characterizing diversity among homologous populations of amplification products. Curr Opin Microbiol 2(3):323–327

Maukonen J, Simoes C, Saarela M (2012) The currently used commercial DNA-extraction methods give different results of clostridial and actinobacterial populations derived from human fecal samples. FEMS Microbiol Ecol 79(3):697–708

Maurice CF, Haiser HJ, Turnbaugh PJ (2013) Xenobiotics shape the physiology and gene expression of the active human gut microbiome. Cell 152(1–2):39–50

McPherson JD (2014) A defining decade in DNA sequencing. Nat Methods 11(10):1003–1005

Miller TL, Wolin MJ (1981) Fermentation by the human large intestine microbial community in an in vitro semicontinuous culture system. Appl Environ Microbiol 42(3):400–407

Minot S, Sinha R, Chen J, Li H, Keilbaugh SA, Wu GD et al (2011) The human gut virome: inter-individual variation and dynamic response to diet. Genome Res 21(10):1616–1625

Musat N, Foster R, Vagner T, Adam B, Kuypers MM (2012) Detecting metabolic activities in single cells, with emphasis on nanoSIMS. FEMS Microbiol Rev 36(2):486–511

Muyzer G, de Waal EC, Uitterlinden AG (1993) Profiling of complex microbial populations by denaturing gradient gel electrophoresis analysis of polymerase chain reaction-amplified genes coding for 16S rRNA. Appl Environ Microbiol 59(3):695–700

Namiki T, Hachiya T, Tanaka H, Sakakibara Y (2012) MetaVelvet: an extension of Velvet assembler to de novo metagenome assembly from short sequence reads. Nucleic Acids Res 40(20), e155

Nguyen TL, Vieira-Silva S, Liston A, Raes J (2015) How informative is the mouse for human gut microbiota research? Dis model mech 8(1):1–16

Nicholson JK, Lindon JC (2008) Systems biology: metabonomics. Nature 455(7216):1054–1056

Nielsen JL, Nielsen PH (2010) Combined microautoradiography and fluorescence *in situ* hybridization (MAR-FISH) for the identification of metabolic active microorgansims. In: Timmis KN (ed) Handbook of hydrocarbon and lilid microbiology. Springer, Berlin, pp 4093–4102

Nielsen HB, Almeida M, Juncker AS, Rasmussen S, Li J, Sunagawa S et al (2014) Identification and assembly of genomes and genetic elements in complex metagenomic samples without using reference genomes. Nat Biotechnol 32(8):822–828

Ott SJ, Musfeldt M, Ullmann U, Hampe J, Schreiber S (2004) Quantification of intestinal bacterial populations by real-time PCR with a universal primer set and minor groove binder probes: a global approach to the enteric flora. J Clin Microbiol 42(6):2566–2572

Paliy O, Agans R (2012) Application of phylogenetic microarrays to interrogation of human microbiota. FEMS Microbiol Ecol 79(1):2–11

Parkhill J, Wren BW (2011) Bacterial epidemiology and biology--lessons from genome sequencing. Genome Biol 12(10):230

Peng Y, Leung HC, Yiu SM, Chin FY (2011) Meta-IDBA: a de Novo assembler for metagenomic data. Bioinformatics 27(13):i94–i101

Pham TA, Lawley TD (2014) Emerging insights on intestinal dysbiosis during bacterial infections. Curr Opin Microbiol 17:67–74

Popa R, Popa R, Mashall MJ, Nguyen H, Tebo BM, Brauer S (2009) Limitations and benefits of ARISA intra-genomic diversity fingerprinting. J Microbiol Methods 78(2):111–118

Porter JR (1976) Antony van Leeuwenhoek: tercentenary of his discovery of bacteria. Bacteriol Rev 40(2):260–269

Prakash T, Taylor TD (2012) Functional assignment of metagenomic data: challenges and applications. Brief Bioinform 13(6):711–727

Quast C, Pruesse E, Yilmaz P, Gerken J, Schweer T, Yarza P et al (2013) The SILVA ribosomal RNA gene database project: improved data processing and web-based tools. Nucleic Acids Res 41(Database issue):D590–D596

Raghunathan A, Ferguson HR Jr, Bornarth CJ, Song W, Driscoll M, Lasken RS (2005) Genomic DNA amplification from a single bacterium. Appl Environ Microbiol 71(6):3342–3347

Rajilic-Stojanovic M, Smidt H, de Vos WM (2007) Diversity of the human gastrointestinal tract microbiota revisited. Environ Microbiol 9(9):2125–2136

Rajilic-Stojanovic M, Heilig HG, Molenaar D, Kajander K, Surakka A, Smidt H et al (2009) Development and application of the human intestinal tract chip, a phylogenetic microarray: analysis of universally conserved phylotypes in the abundant microbiota of young and elderly adults. Environ Microbiol 11(7):1736–1751

Ramirez-Farias C, Slezak K, Fuller Z, Duncan A, Holtrop G, Louis P (2009) Effect of inulin on the human gut microbiota: stimulation of Bifidobacterium adolescentis and Faecalibacterium prausnitzii. Br J Nutr 101(4):541–550

Reardon S (2014) Microbiome therapy gains market traction. Nature 509(7500):269–270

Reck M, Tomasch J, Deng Z, Jarek M, Husemann P, Wagner-Dobler I et al (2015) Stool metatranscriptomics: a technical guideline for mRNA stabilisation and isolation. BMC Genomics 16:494

Reichardt N, Barclay AR, Weaver LT, Morrison DJ (2011) Use of stable isotopes to measure the metabolic activity of the human intestinal microbiota. Appl Environ Microbiol 77(22):8009–8014

Reichardt N, Duncan SH, Young P, Belenguer A, McWilliam Leitch C, Scott KP et al (2014) Phylogenetic distribution of three pathways for propionate production within the human gut microbiota. ISME J 8(6):1323–1335

Reuter JA, Spacek DV, Snyder MP (2015) High-throughput sequencing technologies. Mol Cell 58(4):586–597

Rinke C, Schwientek P, Sczyrba A, Ivanova NN, Anderson IJ, Cheng JF et al (2013) Insights into the phylogeny and coding potential of microbial dark matter. Nature 499(7459):431–437

Robertson M (1980) Biology in the 1980s, plus or minus a decade. Nature 285(5764):358–359

Rogers GB, Cuthbertson L, Hoffman LR, Wing PA, Pope C, Hooftman DA et al (2013) Reducing bias in bacterial community analysis of lower respiratory infections. ISME J 7(4):697–706

Roh SW, Abell GC, Kim KH, Nam YD, Bae JW (2010) Comparing microarrays and next-generation sequencing technologies for microbial ecology research. Trends Biotechnol 28(6):291–299

Salipante SJ, Kawashima T, Rosenthal C, Hoogestraat DR, Cummings LA, Sengupta DJ et al (2014) Performance comparison of Illumina and ion torrent next-generation sequencing platforms for 16S rRNA-based bacterial community profiling. Appl Environ Microbiol 80(24):7583–7591

Salonen A, Lahti L, Salojarvi J, Holtrop G, Korpela K, Duncan SH et al (2014) Impact of diet and individual variation on intestinal microbiota composition and fermentation products in obese men. ISME J 8(11):2218–2230

Salter SJ, Cox MJ, Turek EM, Calus ST, Cookson WO, Moffatt MF et al (2014) Reagent and laboratory contamination can critically impact sequence-based microbiome analyses. BMC Biol 12:87
Sanger F, Nicklen S, Coulson AR (1992) DNA sequencing with chain-terminating inhibitors. 1977 [classical article]. Biotechnology 24:104–108
Savorani F, Rasmussen MA, Mikkelsen MS, Engelsen SB (2013) A primer to nutritional metabolomics by NMR spectroscopy and chemometrics. Food Res Int 54(1):1131–1145
Schloss PD, Westcott SL (2011) Assessing and improving methods used in operational taxonomic unit-based approaches for 16S rRNA gene sequence analysis. Appl Environ Microbiol 77(10):3219–3226
Schloss PD, Westcott SL, Ryabin T, Hall JR, Hartmann M, Hollister EB et al (2009) Introducing mothur: open-source, platform-independent, community-supported software for describing and comparing microbial communities. Appl Environ Microbiol 75(23):7537–7541
Schloss PD, Gevers D, Westcott SL (2011) Reducing the effects of PCR amplification and sequencing artifacts on 16S rRNA-based studies. PLoS One 6(12), e27310
Schuenemann VJ, Singh P, Mendum TA, Krause-Kyora B, Jager G, Bos KI et al (2013) Genome-wide comparison of medieval and modern Mycobacterium leprae. Science 341(6142):179–183
Seedorf H, Griffin NW, Ridaura VK, Reyes A, Cheng J, Rey FE et al (2014) Bacteria from diverse habitats colonize and compete in the mouse gut. Cell 159(2):253–266
Shade A, Hogan CS, Klimowicz AK, Linske M, McManus PS, Handelsman J (2012) Culturing captures members of the soil rare biosphere. Environ Microbiol 14(9):2247–2252
Shanahan F (2015) Separating the microbiome from the hyperbolome. Genome med 7(1):17
Sim K, Cox MJ, Wopereis H, Martin R, Knol J, Li MS et al (2012) Improved detection of bifidobacteria with optimised 16S rRNA-gene based pyrosequencing. PLoS One 7(3), e32543
Sorge RE, Martin LJ, Isbester KA, Sotocinal SG, Rosen S, Tuttle AH et al (2014) Olfactory exposure to males, including men, causes stress and related analgesia in rodents. Nat Methods 11(6):629–632
Stackebrandt E, Ebers J (2011) Taxonomic paramteres revisited: tarenished gold standards. Microbiol Today 33:152–155
Tanca A, Palomba A, Pisanu S, Deligios M, Fraumene C, Manghina V et al (2014) A straightforward and efficient analytical pipeline for metaproteome characterization. Microbiome 2(1):49
Tanner MA, Goebel BM, Dojka MA, Pace NR (1998) Specific ribosomal DNA sequences from diverse environmental settings correlate with experimental contaminants. Appl Environ Microbiol 64(8):3110–3113
Tasse L, Bercovici J, Pizzut-Serin S, Robe P, Tap J, Klopp C et al (2010) Functional metagenomics to mine the human gut microbiome for dietary fiber catabolic enzymes. Genome Res 20(11):1605–1612
Thomas T, Gilbert J, Meyer F (2012) Metagenomics – a guide from sampling to data analysis. Microb inf exp 2(1):3
Tjaden B (2015) De novo assembly of bacterial transcriptomes from RNA-seq data. Genome Biol 16:1
Tottey W, Denonfoux J, Jaziri F, Parisot N, Missaoui M, Hill D et al (2013) The human gut chip "HuGChip", an explorative phylogenetic microarray for determining gut microbiome diversity at family level. PLoS One 8(5), e62544
Tringe SG, Hugenholtz P (2008) A renaissance for the pioneering 16S rRNA gene. Curr Opin Microbiol 11(5):442–446
Tu Q, He Z, Li Y, Chen Y, Deng Y, Lin L et al (2014) Development of HuMiChip for functional profiling of human microbiomes. PLoS One 9(3), e90546
Uchiyama T, Miyazaki K (2009) Functional metagenomics for enzyme discovery: challenges to efficient screening. Curr Opin Biotechnol 20(6):616–622
Uhlik O, Leewis MC, Strejcek M, Musilova L, Mackova M, Leigh MB et al (2013) Stable isotope probing in the metagenomics era: a bridge towards improved bioremediation. Biotechnol Adv 31(2):154–165
Ursell LK, Haiser HJ, Van Treuren W, Garg N, Reddivari L, Vanamala J et al (2014) The intestinal metabolome: an intersection between microbiota and host. Gastroenterology 146(6):1470–1476
Van den Abbeele P, Grootaert C, Marzorati M, Possemiers S, Verstraete W, Gerard P et al (2010) Microbial community development in a dynamic gut model is reproducible, colon region specific, and selective for Bacteroidetes and Clostridium cluster IX. Appl Environ Microbiol 76(15):5237–5246
van Dorst J, Bissett A, Palmer AS, Brown M, Snape I, Stark JS et al (2014) Community fingerprinting in a sequencing world. FEMS Microbiol Ecol 89(2):316–330
van Opijnen T, Camilli A (2013) Transposon insertion sequencing: a new tool for systems-level analysis of microorganisms. Nat Rev Microbiol 11(7):435–442
Verberkmoes NC, Russell AL, Shah M, Godzik A, Rosenquist M, Halfvarson J et al (2009) Shotgun metaproteomics of the human distal gut microbiota. ISME J 3(2):179–189
Vetrovsky T, Baldrian P (2013) The variability of the 16S rRNA gene in bacterial genomes and its consequences for bacterial community analyses. PLoS One 8(2), e57923
Wagner M (2009) Single-cell ecophysiology of microbes as revealed by Raman microspectroscopy or secondary ion mass spectrometry imaging. Annu Rev Microbiol 63:411–429
Walker A (2014) Adding genomic 'foliage' to the tree of life. Nat Rev Microbiol 12(2):78
Walker A, Parkhill J (2008) Single-cell genomics. Nat Rev Microbiol 6(3):176–177

Walker AW, Duncan SH, Louis P, Flint HJ (2014) Phylogeny, culturing, and metagenomics of the human gut microbiota. Trends Microbiol 22(5):267–274

Walker AW, Martin JC, Scott P, Parkhill J, Flint HJ, Scott KP (2015) 16S rRNA gene-based profiling of the human infant gut microbiota is strongly influenced by sample processing and PCR primer choice. Microbiome 3:26

Wesolowska-Andersen A, Bahl MI, Carvalho V, Kristiansen K, Sicheritz-Ponten T, Gupta R et al (2014) Choice of bacterial DNA extraction method from fecal material influences community structure as evaluated by metagenomic analysis. Microbiome 2:19

Wikoff WR, Anfora AT, Liu J, Schultz PG, Lesley SA, Peters EC et al (2009) Metabolomics analysis reveals large effects of gut microflora on mammalian blood metabolites. Proc Natl Acad Sci U S A 106(10):3698–3703

Wilmes P, Bond PL (2009) Microbial community proteomics: elucidating the catalysts and metabolic mechanisms that drive the Earth's biogeochemical cycles. Curr Opin Microbiol 12(3):310–317

Woese CR, Fox GE (1977) Phylogenetic structure of the prokaryotic domain: the primary kingdoms. Proc Natl Acad Sci U S A 74(11):5088–5090

# The Gut Microbiota and their Metabolites: Potential Implications for the Host Epigenome

3

Mona Mischke and Torsten Plösch

**Abstract**

The gut microbiota represents a metabolically active biomass of up to 2 kg in adult humans. Microbiota-derived molecules significantly contribute to the host metabolism. Large amounts of bacterial metabolites are taken up by the host and are subsequently utilized by the human body. For instance, short chain fatty acids produced by the gut microbiota are a major energy source of humans.

It is widely accepted that microbiota-derived metabolites are used as fuel for beta-oxidation (short chain fatty acids) and participate in many metabolic processes (vitamins, such as folic acid). Apart from these direct metabolic effects, it also becomes more and more evident that these metabolites can interact with the mammalian epigenetic machinery. By interacting with histones and DNA they may be able to manipulate the host's chromatin state and functionality and hence its physiology and health.

In this chapter, we summarize the current knowledge on possible interactions of different bacterial metabolites with the mammalian epigenetic machinery, mostly based on in vitro data. We discuss the putative impact on chromatin marks, for example histone modifications and DNA methylation. Subsequently, we speculate about possible beneficial and adverse consequences for the epigenome, the physiology and health of the host, as well as plausible future applications of this knowledge for in vivo translation to support personal health.

M. Mischke
Nutricia Research,
Uppsalalaan 12, 3584 CT Utrecht, The Netherlands
e-mail: mona.mischke@danone.com

T. Plösch (✉)
Department of Obstetrics and Gynaecology,
University Medical Center Groningen,
Hanzeplein 1, 9700 RB Groningen, The Netherlands
e-mail: t.plosch@umcg.nl;
http://www.epigenetic-programming.nl



A. Schwiertz (ed.), *Microbiota of the Human Body*, Advances in Experimental Medicine and Biology 902, DOI 10.1007/978-3-319-31248-4_3

**Keywords**
Bacterial metabolites • Epigenetics • (early life) programming • (early life) nutrition • DNA methylation • Histone modification • Breast feeding

## 3.1 Introduction

It has been known for a long time that the mammalian gut microbiota is indispensable for the production of vitamins and the fermentation of otherwise indigestible components of the diet (Albert et al. 1980; Cummings and Macfarlane 1997; Ramotar et al. 1984). In ruminants for example, the microbiota produces the majority of fatty acids used by the organism as an energy source, and also humans make good use of the energy and specific metabolites produced by the gut microbiota (Cook and Sellin 1998).

In the last decade it became clear that the microbiota is not only a producer of nutrients and vitamins, but also closely interacts with the host. Experiments in rodents and humans have shown that the composition of the gut microbiota is related to the nutritional, but also to the metabolic and health status of the host (David et al. 2014; Ridaura et al. 2013; Tremaroli and Backhed 2012; Remely et al. 2013; Le Chatelier et al. 2013). So, it was observed that the metabolic adaption to a high-fat diet is associated with changes in the gut microbiota in mice (Serino et al. 2012). Moreover, the gut microbial phylotypes in hundred different inbred mouse strains correlate with the level of obesity in response to a high-fat/high-sucrose diet (Parks et al. 2013), and the status of gut colonization is linked with immune response and longevity (Tazume et al. 1991; Round and Mazmanian 2009). In humans, similar associations between diet, microbiota, and health were detected (Tremaroli and Backhed 2012; Claesson et al. 2012). Several studies link the composition of the gut microbiota to obesity in humans, although their partly conflicting results highlight that the exact role of the gut microbiota and causality remains to be elucidated (Duncan et al. 2008; Jumpertz et al. 2011; Ley et al. 2006; Schwiertz et al. 2010; Zhang et al. 2009). Nevertheless, one of the most prominent studies in humans was the transfer of the gut microbiota from lean to obese volunteers with insulin resistance that resulted in restored insulin sensitivity in the recipients (Vrieze et al. 2012). The obvious control experiment – transfer from obese to lean volunteers – could not be performed because of ethical reasons.

These and similar studies show that the gut microbiota plays an important role in our health and survival. One relevant link between nutrition, microbiota, and health outcome that has to be explored in more detail, is the metabolite mix produced by the microbiota, the so called metabolome, which forms a specific entity together with the microbiota (McHardy et al. 2013). Specific microbiota-mediated metabolite profiles can be associated with the predisposition to metabolic impairments, such as impaired glucose homeostasis and non-alcoholic fatty liver disease (NAFLD) (Dumas et al. 2006). Very recently, we have put forward the hypothesis that the gut microbiota by means of the associated metabolome may influence the host's long-term physiology via modulating its epigenome (Mischke and Plosch 2013). Here, we will extend and underline this notion on the basis of reviewing a number of relevant bacterial metabolites and their documented effects.

## 3.2 Epigenetics - the Mediator Between the Genome and Physiology

Functionally, epigenetics explains how a multicellular, differentiated organism can be constructed from a uniform genome. This implies that during differentiation the activity of genes can be specifically increased or silenced depending on whether or not they are needed in a certain cell type – and to which extent. This regulation basically takes place on the level of DNA modification and DNA accessibility, achieved by the close interaction of DNA methylation and histone modifications, which form the epigenome (Cedar and Bergman 2009; Bernstein et al. 2007).

On the level of DNA, the activity of a gene can be regulated by specific methylation of CpG dinucleotides in a certain gene region. DNA methylation of gene promoters or enhancers often blocks the transcriptional activity of the appertaining gene (Jaenisch and Bird 2003). In general, the combined methylation of multiple CpG dinucleotides contributes to the resulting activity state. Once methylated, the methylation is in principle stable including subsequent cell generations. This is achieved by faithful maintenance methylation of the new DNA strands formed during semiconservative replication, according to the template DNA strand (Sharif et al. 2007; Hermann et al. 2004). Thus, DNA methylation of specific CpG positions can be fully conserved and therefore carried over to the daughter cells as transcriptional label. Removal of the DNA methylation label requires either DNA synthesis without sufficient maintenance methylation, which gradually dilutes the DNA methylation marks, or selective oxidation of methylcytosine to hydroxymethylcytosine (Wu and Zhang 2014).

The accessibility of DNA for transcription is mainly regulated by covalent modifications at specific amino acid residues of the histone tails. These modifications determine the steric conformation of the histones and therefore the packaging density of the chromatin. On the contrary to DNA methylation, where only a few enzymes are involved in methylation and de-methylation of the DNA, the situation for histone modification is far more complex: numerous modifications of the different histones are known, including but not limited to acetylation, methylation, and phosphorylation of different amino acid residues. This requires a plethora of enzymes from several classes to attach and remove the functional groups to and from the amino acid residues of the histone tails (Plass et al. 2013; http://www.cellsignal.com/contents/resources-reference-tables/histone-modification-table/science-tables-histone). Ultimately, DNA methylation and histone modifications often closely interact with each other, together determining the transcriptional activity of a locus and shaping the epigenetic landscape (Cedar and Bergman 2009).

## 3.3 Interaction of Bacterial Metabolites and the Host's Metabolic Regulation

Obviously, both DNA methylation and histone modifications require substrates. For example, the availability of C1 metabolites is necessary for methylation reactions as is ATP for phosphorylation (Jimenez-Chillaron et al. 2012). Although it is not plausible that such important regulatory reactions are exclusively regulated on the level of substrates, severe or long-lasting fluctuations in substrate availability may influence the level of modifications. Such fluctuations in substrate availability may occur in extraordinary physiological situations, including undernutrition during pregnancy, as seen in the Dutch Hunger Winter cohort. In this study cohort, DNA hypomethylation of some genes was observed, which may partially be related to substrate availability. However, in the same persons other genes have been shown to be hypermethylated, which argues against one single mechanism of action (Heijmans et al. 2008). Moreover, less extraordinary, more naturally occurring variations in substrate availability, as for example seen during the cycle of the seasons or due to regional dietary preferences, were also identified to have a specific impact on epigenetic modifications (Waterland et al. 2010).

Numerous epigenetically relevant substances are not only provided by the diet, but are also produced by the gut bacteria. Thus, apart from direct nutritional variation as described above, also changes in the gut microbiota have the potential to contribute to substrate fluctuations through the associated metabolome. Eventually, not only epigenetically active substances from the diet, but also significant levels of epigenetically active bacterial metabolites reach cells of the host organism (Table 3.1). For example, C1 groups for methylation reactions are provided by the

**Table 3.1** Bacterial metabolites that are indicated to have an epigenetic function

| Metabolite | Model | Epigenetic action |
|---|---|---|
| Short-chain fatty acids (SCFA), general | In vitro, human | HDAC inhibition (Waldecker et al. 2008) |
| | | Associated to LINE-1 DNA methylation (Worthley et al. 2011) |
| Acetate (C2:0) | In vitro | HDAC inhibition; histone (H3, H4) hyper-acetylation (Sealy and Chalkley 1978) |
| Propionate (C3:0) | In vitro | HDAC inhibition; histone (H3, H4) hyper-acetylation (Sealy and Chalkley 1978; Takenaga 1986) |
| Butyrate (C4:0) | In vitro, in vivo | HDAC inhibition of HDAC class I, IIa, and IV (Davie 2003) |
| | | Regulation of transcription factor availability (Blottiere et al. 2003) |
| Valerate (C5:0) | In vitro | HDAC inhibition (Ortiz-Caro et al. 1986) |
| Branched-chain fatty acids (BCFA), general | | |
| Isobutyrate | In vitro | HDAC inhibition (Waldecker et al. 2008) |
| | | Increased histone acetylation, probably via HDAC inhibition (Suzuki-Mizushima et al. 2002) |
| Isovalerate | In vitro | HDAC inhibition (Waldecker et al. 2008) |
| | | Increased histone acetylation, probably via HDAC inhibition (Suzuki-Mizushima et al. 2002) |
| Organic acids, general | In vitro, in vivo | HDAC inhibition by low pH (Latham et al. 2012) |
| Lactate (D-, L-lactate) | In vitro, in vivo | (weak) HDAC inhibition (Latham et al. 2012) |
| Phenolic compounds | In vitro, in vivo | HDAC inhibition (Waldecker et al. 2008) |
| | | Bacterial break down of dietary polyphenols (quercetin, curcumin, catechin) redering them unavailable for affecting HDAC activity (Rajendran et al. 2011) |
| Phenylbutyrate | In vitro | HDAC inhibition (Lea and Tulsyan 1995; Lea et al. 2004) |
| Phenylacetate | In vitro | HDAC inhibition (Waldecker et al. 2008; Lea and Tulsyan 1995) |
| | | Reduction of reactive oxygen species, which otherwise affect HAT and HDAC activity and increase DNA methylation (Beloborodova et al. 2012) |
| 4-hydroxyphenylacetate | In vitro | HDAC inhibition (Waldecker et al. 2008) |
| | | Reduction of reactive oxygen species, which otherwise affect HAT and HDAC activity and increase DNA methylation (Beloborodova et al. 2012) |
| Phenylpropionate | In vitro | HDAC inhibition (Waldecker et al. 2008) |
| 4-hydroxyphenylpropionate | In vitro | HDAC inhibition (Waldecker et al. 2008) |

(continued)

**Table 3.1** (continued)

| Metabolite | Model | Epigenetic action |
|---|---|---|
| p-cresol | In vitro, mouse | Expression induction of DNA methyltransferases 1, 3a, and 3b (Sun et al. 2012) |
| | | CpG hypermethylation of Klotho gene; decreased Klotho expression (Sun et al. 2012) |
| Sulfur compounds | In vitro, in vivo | Histone modifications (Canani et al. 2011) |
| Hydrogen sulfide (H2S) | In vitro, rat | Inhibition of cell proliferation by epigenetic mechanism reducing recruitment of Brg1 to relevant promoter regions (Li et al. 2013) |
| | | Reduction/neutralization of reactive oxygen species, which otherwise affect HAT and HDAC activity and increase DNA methylation (Afanas'ev 2014) |
| Cell wall components | | |
| Lipopolysaccharide (LPS) | In vitro | Chromatin modification at IL-8 gene, including histone H3 acetylation and methylation; IL-8 activation (Angrisano et al. 2010) |
| Peptidoglycan (PGN) | In vitro | Modulation of chromatin structure and transcriptional activity at Foxp3 locus (Lal et al. 2011) |
| Lipoteichoic acid (LTA) | Mouse, in vitro | Potential epigenetic regulation of genes in colorectal cancer (Lightfoot et al. 2013) |
| Vitamins | | |
| Thiamine (vitamin B1) | In vitro, in vivo | As coenzyme involved in generation of ATP; critical for phosphorylation reactions (Hill 1997) |
| Riboflavin (vitamin B2) | In vitro, in vivo | As cofactor involved in one-carbone metabolism; critical for methylation reactions (Anderson et al. 2012) |
| Niacin (vitamin B3) | In vitro, in silico | SIRT (Class III HDAC) inhibitor (Avalos et al. 2005; Denu 2005) |
| Pantothenate (vitamin B5) | In vitro, in vivo | As coenzyme A substrate for acylation and acetylation reactions; signal transduction, enzyme activity regulation (Marmorstein 2001) |
| | | Histone hypoacetylation and fragile DNA by impaired Coenzyme A synthesis (Cai et al. 2011) |
| Pyridoxine (vitamin B6) | In vitro, in vivo | As cofactor involved in one-carbone metabolism; critical for methylation reactions (Anderson et al. 2012) |
| Biotin (vitamin B7) | In vitro, drosophila | Substrate for histone biotinylation; gene activity regulation and transposable element repression (Hassan and Zempleni 2008) |
| | | Decreased histone biotinylation associated with life span and stress resistance in Drosophila (Zempleni et al. 2008) |

(continued)

**Table 3.1** (continued)

| Metabolite | Model | Epigenetic action |
|---|---|---|
| Folate (vitamin B9) | In vitro, in vivo, rat | As methyl donor involved in one-carbone metabolism; critical for methylation reaction (Kalhan 2013; Anderson et al. 2012) |
| | | Activity reduction of DNA methyltransferase (Ly et al. 2011) |
| Cobalamin (vitamin B12) | In vitro, in vivo | As cofactor involved in one-carbone metabolism; critical for methylation reactions (Kalhan 2013; Anderson et al. 2012) |
| Choline | In vitro, mouse | Methyl donor; loses availability through break-down by human gut microbiota (Dumas et al. 2006) |
| | | DNA methylation and gene expression changes in colitis (Schaible et al. 2011) |
| | | Broken down to TMAO and betaine (Wang et al. 2011) |
| Betaine | In vitro, in vivo, human | As methyl donor involved in one-carbone metabolism; critical for methylation reaction (Kalhan 2013; Canani et al. 2011) |
| | | Alterations of DNA methylation associated with abnormalities of DNA methyltransferases in human cancers (Kanai and Hirohashi 2007) |
| Trimethylamin-N-oxid (TMAO) | In vitro, mouse | Break down product of methyl donor choline (Wang et al. 2011) |
| Equol | In vitro | Ppromoter CpG island hypomethylation of breast cancer susceptibility genes BRCA1 and BRCA2; increase of BRCA1 and BRCA2 proteins (Bosviel et al. 2012) |
| Ammonium ($NH_4$) | Human | Inverse association of fecal $NH_4$ and rectal LINE-1 methylation (Worthley et al. 2011) |
| α-ketoglutarate | In vitro, human | Effect on histone and DNA (de)methylation; co-factor of HDM and TET protein family memebers (Wang et al. 2013; Hou and Yu 2010) |
| Conjugated linoleic acids (CLA) | In vitro, in vivo | Increased SIRT1 deacetylation activity by trans-10, cis-12 CLA treatment via reciprocal activation of AMPK (Jiang et al. 2012) |
| | | Decreased histone phosphorylation by CDK2 inhibition (Cho et al. 2006) |

*HDM* histone demethylase, *TET* Ten-eleven translocation protein

one-carbon metabolism in the form of the co-substrate S-adenosyl methionine (Dominguez-Salas et al. 2013). Known bacterial metabolites contributing to a well-functioning one-carbon metabolism include various vitamins and their metabolites, such as folate (vitamin B9), cobalamin (vitamin B12), pyridoxine (vitamin B6), riboflavin (vitamine B2), and betaine (Kalhan 2013; Anderson et al. 2012). Dietary choline, on the other hand, which contributes as methyl donor to the one-carbon metabolism, can be metabolized by bacteria to trimethylamin-N-oxid (TMAO) and betaine, effectively rendering this methyl donor unavailable for the host (Schaible et al. 2011; Wang et al. 2011, 2014). Other bacterial metabolites that are known to function as substrates for epigenetic modifications are biotin (vitamin B7) for biotinylation of histones (Hassan and Zempleni 2008; Zempleni et al. 2008) and pantothenic acid (vitamin B5) to form acetyl-CoA for acetylation reactions (Marmorstein 2001; Nakamura et al. 2012; Cai et al. 2011).

Another level of complexity is added by the fact that several metabolites produced by the gut microbiota can influence epigenetic regulation, for example by inhibiting modifying enzymes. The most prominent example is short chain fatty acids, which are known to inhibit mammalian histone deacetylases (HDAC). Extensive in vitro studies have shown that HDAC inhibition by these short chain fatty acids links to an increase in the average histone acetylation level of the cell, in turn affecting gene activity (Blottiere et al. 2003; Davie 2003; Ortiz-Caro et al. 1986; Sealy and Chalkley 1978; Takenaga 1986; Waldecker et al. 2008). In this context, butyrate is considered the most active HDAC inhibitor, followed by propionate (Table 3.1). Remarkably, both butyrate and propionate are highly prominent in the gut (Cook and Sellin 1998), and their abundance is influenced among others by the host's diet and the bacterial population present (Cummings 1984; Cummings and Englyst 1987; McBurney and Thompson 1990). Apart from short chain fatty acids, also other bacterial metabolites have been indicated to influence the activity of the epigenetic machinery, either directly or indirectly: organic acids such as lactate are known to be weak HDAC inhibitors via lowering the pH (Latham et al. 2012); specific phenolic and sulfur compounds are indicated to impact DNA methylation and histone acetylation via influencing cellular levels of reactive oxygen species (Afanas'ev 2014; Li et al. 2013; Beloborodova et al. 2012); and niacin (vitamin B3) as well as conjugated linoleic acids affect the activity of sirtuins, which are class III HDACs (Avalos et al. 2005; Denu 2005). The majority of these examples relates to histone modifications, but there are also a few indications that the activity of DNA methylating enzymes can be regulated similarly by metabolites, such as folate, betaine, α-ketoglutarate, or p-cresol, which can be products of the bacterial metabolism (Sun et al. 2012; Ly et al. 2011; Kanai and Hirohashi 2007; Wang et al. 2013).

## 3.4 Putative Targets of Bacterial Metabolites - in Time and Space

As for metabolites of the gut microbiota, the cells of the intestinal epithelium, including their stem cells, are most likely to be reached by epigenetically relevant concentrations of these metabolites. In the second line, immunocompetent cells that are in proximity of the gut, but also the liver are putatively exposed to these metabolites. It is therefore conceivable that the epigenome of these cells may be influenced by different fluxes of bacterial metabolites under different metabolic circumstances. However, for most of the metabolites discussed the fluxes have not yet specifically been measured.

At least for DNA methylation it is known that the organism is especially vulnerable during critical windows in development, including the intrauterine and early post-natal life, which refers to the "first 1,000 days of life" concept (Arenz et al. 2004; Zhang et al. 2012). One of the underlying reasons may be the fact that this is a period where dividing cells require relative high levels of methyl donors to methylate CpG positions at the newly formed DNA strand to maintain the established methylation patterns. In the absence of a sufficient supply of methyl donors, methyl marks may then be purely lost by dilution (Wu and Zhang 2014).

During pre- and postnatal development the body is formed in an orchestrated sequence of steps, forming different cell types and tissues. Epigenetically vulnerable periods vary for the different organs. Organs like the heart which are formed early in embryogenesis, when exposure to bacterial metabolites occurs only indirectly via the mother, may therefore be almost inert against these factors. On the other hand, organs being formed and differentiated in a later period, like the brain, or with high cell division rates, like the intestine, might be more susceptible to the influence of metabolites from the microbiota. This may explain the relationship between timing of the stimulus and the observed physiological outcome (Rueda-Clausen et al. 2011). Another wave of epigenetic events may then occur during puberty, when extensive remodeling processes take place, possibly linked with epigenomic re-arrangements.

Apart from certain developmental stages that require sufficient abundance of substrates, such as methyl donors, to satisfy increased substrate expenditure, it should be noted that also situations of unbalanced nutrient supply might lead to a *relative* substrate shortage with associated effects on the epigenome. It is conceivable that in situations with a low intake of micronutrients and high intake of fat and proteins, as often seen in obesity, the delicate balance between supply and demand of necessary substances is most likely disturbed. On the one hand this could be due to direct effects of low quality diets and higher requirements of a bigger body volume, on the other hand specific dietary effects on the microbiota and its metabolome might contribute to this.

## 3.5 Health Implications - Sooner or Later

We have recently proposed that infant nutrition may have long-term physiological consequences, mediated by the epigenome (Mischke and Plosch 2013). We speculated that early-life nutrition may determine the composition of the early gut microbiota and therefore the absolute amounts and ratios of bacterial metabolites, including butyrate and folate. As a consequence, tissues exposed to high concentrations of these metabolites, like the intestinal epithelium or the liver, may undergo epigenetic changes, which might program their long-term metabolic set points and regulation. This, in turn, may alter organ plasticity linked to the health status of the host. Obviously, this hypothesis needs to be thoroughly tested.

One possibility to test if early-life influences on the microbiota might affect the host's long-term health via associated changes of the microbiota's metabolome and the host's epigenome is the exploration of the metabolic and epigenetic consequences of Caesarean sections. Future studies should specifically address how the well-characterized differences in the microbiota upon Caesarean sections (Dominguez-Bello et al. 2010) affect the microbial metabolome and epigenome of the host.

One of the most relevant health problems in the twenty-first century is the increasing incidence and severity of obesity. It has been shown in animal experiments, but also clinical studies, that the microbiota of lean and obese subjects differs significantly. Moreover, manipulation of the microbiota has been shown to improve metabolic parameters as well as body weight directly, leading to the idea that the gut microbiota could pose a novel target for weight control. Specifically, the first human trials of fecal transplants from lean to obese people initiated a debate whether this technology might be a novel treatment strategy for obesity. This adds to the long lasting discussion over possible dietary or supplementary interventions with pre- and probiotics to improve human health.

On the background of our ideas, the situation gets even more complex. Bacterial metabolites, impacting the host epigenome, could not only have direct but especially long-lasting physiological and health consequences. For any intervention aimed at the gut microbiota, this would imply that changing the microbiota might initiate long-lasting effects, which need to be considered before these interventions can be viewed as safe. On the other hand, (early-life) microbiota-related strategies may not only have tremendous impact, but consequently also offers enormous opportunities regarding prevention of disease and modulation of long-term health. A first, rather simple step to neutralize negative effects of the delivery mode on microbiome, metabolome and epigenome of the infant might be the inoculation of infants after Caesarean section with a natural bacterial population. Beyond this, it may be

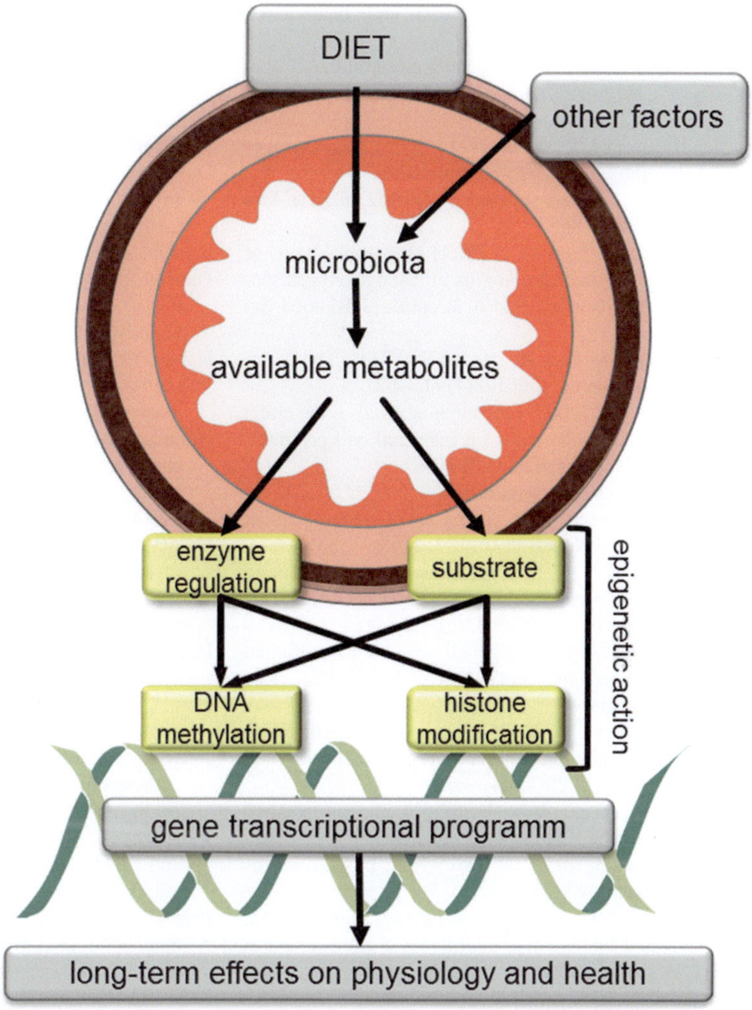


**Fig. 3.1** Proposed mechanism how epigenetic actions that potentially impact long-term physiology and health are linked to (early-life) factors via the microbiota and its metabolites (Modified from (Mischke and Plosch 2013))

feasible one day to integrate our knowledge of the host-microbiota interaction into novel nutritional strategies for infants. Apart from beneficially affecting bacterial conditions for optimal long-term health in general, these strategies could also be aimed at specific risk groups, such as infants born after severe pregnancy complications, to counteract adverse effects. Therefore, future developments might lead to approaches of targeted microbiota management – as preventive or therapeutic strategy – to support personal health (Fig. 3.1).

**Acknowledgments** We thank Dr. Eline van der Beek, Sebastian Tims, Dr. Jan Knol and Dr. Sicco Scherjon for helpful comments and stimulating discussions and Charlotte Snethla for supporting the data compilation.

## References

Afanas'ev I (2014) New nucleophilic mechanisms of ros-dependent epigenetic modifications: comparison of aging and cancer. Aging Dis 5(1):52–62

Albert MJ, Mathan VI, Baker SJ (1980) Vitamin B12 synthesis by human small intestinal bacteria. Nature 283(5749):781–782

Anderson OS, Sant KE, Dolinoy DC (2012) Nutrition and epigenetics: an interplay of dietary methyl donors, one-carbon metabolism and DNA methylation. J Nutr Biochem 23(8):853–859

Angrisano T et al (2010) LPS-induced IL-8 activation in human intestinal epithelial cells is accompanied by specific histone H3 acetylation and methylation changes. BMC Microbiol 10:172

Arenz S et al (2004) Breast-feeding and childhood obesity – a systematic review. Int J Obes Relat Metab Disord 28(10):1247–1256

Avalos JL, Bever KM, Wolberger C (2005) Mechanism of sirtuin inhibition by nicotinamide: altering the NAD(+) cosubstrate specificity of a Sir2 enzyme. Mol Cell 17(6):855–868

Beloborodova N et al (2012) Effect of phenolic acids of microbial origin on production of reactive oxygen species in mitochondria and neutrophils. J Biomed Sci 19:89

Bernstein BE, Meissner A, Lander ES (2007) The mammalian epigenome. Cell 128(4):669–681

Blottiere HM et al (2003) Molecular analysis of the effect of short-chain fatty acids on intestinal cell proliferation. Proc Nutr Soc 62(1):101–106

Bosviel R et al (2012) Epigenetic modulation of BRCA1 and BRCA2 gene expression by equol in breast cancer cell lines. Br J Nutr 108(7):1187–1193

Cai L et al (2011) Acetyl-CoA induces cell growth and proliferation by promoting the acetylation of histones at growth genes. Mol Cell 42(4):426–437

Canani RB et al (2011) Epigenetic mechanisms elicited by nutrition in early life. Nutr Res Rev 24(2):198–205

Cedar H, Bergman Y (2009) Linking DNA methylation and histone modification: patterns and paradigms. Nat Rev Genet 10(5):295–304

Cell Signaling Technology. Histone modification table. Available from: http://www.cellsignal.com/contents/resources-reference-tables/histone-modification-table/science-tables-histone

Cho HJ et al (2006) Trans-10, cis-12, not cis-9, trans-11, conjugated linoleic acid inhibits G1-S progression in HT-29 human colon cancer cells. J Nutr 136(4):893–898

Claesson MJ et al (2012) Gut microbiota composition correlates with diet and health in the elderly. Nature 488(7410):178–184

Cook SI, Sellin JH (1998) Review article: short chain fatty acids in health and disease. Aliment Pharmacol Ther 12(6):499–507

Cummings JH (1984) Microbial digestion of complex carbohydrates in man. Proc Nutr Soc 43(1):35–44

Cummings JH, Englyst HN (1987) Fermentation in the human large intestine and the available substrates. Am J Clin Nutr 45(5 Suppl):1243–1255

Cummings JH, Macfarlane GT (1997) Colonic microflora: nutrition and health. Nutrition 13(5):476–478

David LA et al (2014) Diet rapidly and reproducibly alters the human gut microbiome. Nature 505(7484):559–563

Davie JR (2003) Inhibition of histone deacetylase activity by butyrate. J Nutr 133(7 Suppl):2485S–2493S

Denu JM (2005) Vitamin B3 and sirtuin function. Trends Biochem Sci 30(9):479–483

Dominguez-Bello MG et al (2010) Delivery mode shapes the acquisition and structure of the initial microbiota across multiple body habitats in newborns. Proc Natl Acad Sci U S A 107(26):11971–11975

Dominguez-Salas P et al (2013) DNA methylation potential: dietary intake and blood concentrations of one-carbon metabolites and cofactors in rural African women. Am J Clin Nutr 97(6):1217–1227

Dumas ME et al (2006) Metabolic profiling reveals a contribution of gut microbiota to fatty liver phenotype in insulin-resistant mice. Proc Natl Acad Sci U S A 103(33):12511–12516

Duncan SH et al (2008) Human colonic microbiota associated with diet, obesity and weight loss. Int J Obes (Lond) 32(11):1720–1724

Hassan YI, Zempleni J (2008) A novel, enigmatic histone modification: biotinylation of histones by holocarboxylase synthetase. Nutr Rev 66(12):721–725

Heijmans BT et al (2008) Persistent epigenetic differences associated with prenatal exposure to famine in humans. Proc Natl Acad Sci U S A 105(44):17046–17049

Hermann A, Goyal R, Jeltsch A (2004) The Dnmt1 DNA-(cytosine-C5)-methyltransferase methylates DNA processively with high preference for hemimethylated target sites. J Biol Chem 279(46):48350–48359

Hill MJ (1997) Intestinal flora and endogenous vitamin synthesis. Eur J Cancer Prev 6(Suppl 1):S43–S45

Hou H, Yu H (2010) Structural insights into histone lysine demethylation. Curr Opin Struct Biol 20(6):739–748

Jaenisch R, Bird A (2003) Epigenetic regulation of gene expression: how the genome integrates intrinsic and environmental signals. Nat Genet 33(Suppl):245–254

Jiang S et al (2012) Cross regulation of sirtuin 1, AMPK, and PPARgamma in conjugated linoleic acid treated adipocytes. PLoS ONE 7(11), e48874

Jimenez-Chillaron JC et al (2012) The role of nutrition on epigenetic modifications and their implications on health. Biochimie 94(11):2242–2263

Jumpertz R et al (2011) Energy-balance studies reveal associations between gut microbes, caloric load, and nutrient absorption in humans. Am J Clin Nutr 94(1):58–65

Kalhan SC (2013) One-carbon metabolism, fetal growth and long-term consequences. Nestle Nutr Inst Work Ser 74:127–138

Kanai Y, Hirohashi S (2007) Alterations of DNA methylation associated with abnormalities of DNA methyltransferases in human cancers during transition from a precancerous to a malignant state. Carcinogenesis 28(12):2434–2442

Lal G et al (2011) Distinct inflammatory signals have physiologically divergent effects on epigenetic regulation of Foxp3 expression and Treg function. Am J Transplant 11(2):203–214

Latham T et al (2012) Lactate, a product of glycolytic metabolism, inhibits histone deacetylase activity and promotes changes in gene expression. Nucleic Acids Res 40(11):4794–4803

Le Chatelier E et al (2013) Richness of human gut microbiome correlates with metabolic markers. Nature 500(7464):541–546

Lea MA, Tulsyan N (1995) Discordant effects of butyrate analogues on erythroleukemia cell proliferation, differentiation and histone deacetylase. Anticancer Res 15(3):879–883

Lea MA et al (2004) Induction of histone acetylation and inhibition of growth by phenyl alkanoic acids and structurally related molecules. Cancer Chemother Pharmacol 54(1):57–63

Ley RE et al (2006) Microbial ecology: human gut microbes associated with obesity. Nature 444(7122):1022–1023

Li L et al (2013) Brg1-dependent epigenetic control of vascular smooth muscle cell proliferation by hydrogen sulfide. Biochim Biophys Acta 1833(6):1347–1355

Lightfoot YL et al (2013) Targeting aberrant colon cancer-specific DNA methylation with lipoteichoic acid-deficient Lactobacillus acidophilus. Gut Microbes 4(1):84–88

Ly A et al (2011) Effect of maternal and postweaning folic acid supplementation on mammary tumor risk in the offspring. Cancer Res 71(3):988–997

Marmorstein R (2001) Protein modules that manipulate histone tails for chromatin regulation. Nat Rev Mol Cell Biol 2(6):422–432

McBurney MI, Thompson LU (1990) Fermentative characteristics of cereal brans and vegetable fibers. Nutr Cancer 13(4):271–280

McHardy IH et al (2013) Integrative analysis of the microbiome and metabolome of the human intestinal mucosal surface reveals exquisite inter-relationships. Microbiomed 1(1):17

Mischke M, Plosch T (2013) More than just a gut instinct-the potential interplay between a baby's nutrition, its gut microbiome, and the epigenome. Am J Physiol Regul Integr Comp Physiol 304(12):R1065–R1069

Nakamura T et al (2012) Impaired coenzyme A synthesis in fission yeast causes defective mitosis, quiescence-exit failure, histone hypoacetylation and fragile DNA. Open Biol 2(9):120117

Ortiz-Caro J et al (1986) Modulation of thyroid hormone nuclear receptors by short-chain fatty acids in glial C6 cells. Role of histone acetylation. J Biol Chem 261(30):13997–14004

Parks BW et al (2013) Genetic control of obesity and gut microbiota composition in response to high-fat, high-sucrose diet in mice. Cell Metab 17(1):141–152

Plass C et al (2013) Mutations in regulators of the epigenome and their connections to global chromatin patterns in cancer. Nat Rev Genet 14(11):765–780

Rajendran P et al (2011) Metabolism as a key to histone deacetylase inhibition. Crit Rev Biochem Mol Biol 46(3):181–199

Ramotar K et al (1984) Production of menaquinones by intestinal anaerobes. J Infect Dis 150(2):213–218

Remely M et al (2013) Effects of short chain fatty acid producing bacteria on epigenetic regulation of FFAR3 in type 2 diabetes and obesity. Gene 537(1):85–92

Ridaura VK et al (2013) Gut microbiota from twins discordant for obesity modulate metabolism in mice. Science 341(6150):1241214

Round JL, Mazmanian SK (2009) The gut microbiota shapes intestinal immune responses during health and disease. Nat Rev Immunol 9(5):313–323

Rueda-Clausen CF, Morton JS, Davidge ST (2011) The early origins of cardiovascular health and disease: who, when, and how. Semin Reprod Med 29(3):197–210

Schaible TD et al (2011) Maternal methyl-donor supplementation induces prolonged murine offspring colitis susceptibility in association with mucosal epigenetic and microbiomic changes. Hum Mol Genet 20(9):1687–1696

Schwiertz A et al (2010) Microbiota and SCFA in lean and overweight healthy subjects. Obesity (Silver Spring) 18(1):190–195

Sealy L, Chalkley R (1978) The effect of sodium butyrate on histone modification. Cell 14(1):115–121

Serino M et al (2012) Metabolic adaptation to a high-fat diet is associated with a change in the gut microbiota. Gut 61(4):543–553

Sharif J et al (2007) The SRA protein Np95 mediates epigenetic inheritance by recruiting Dnmt1 to methylated DNA. Nature 450(7171):908–912

Sun CY, Chang SC, Wu MS (2012) Suppression of Klotho expression by protein-bound uremic toxins is associated with increased DNA methyltransferase expression and DNA hypermethylation. Kidney Int 81(7):640–650

Suzuki-Mizushima Y et al (2002) Enhancement of NGF- and cholera toxin-induced neurite outgrowth by butyrate in PC12 cells. Brain Res 951(2):209–217

Takenaga K (1986) Effect of butyric acid on lung-colonizing ability of cloned low-metastatic Lewis lung carcinoma cells. Cancer Res 46(3):1244–1249

Tazume S et al (1991) Effects of germfree status and food restriction on longevity and growth of mice. Jikken Dobutsu 40(4):517–522

Tremaroli V, Backhed F (2012) Functional interactions between the gut microbiota and host metabolism. Nature 489(7415):242–249

Vrieze A et al (2012) Transfer of intestinal microbiota from lean donors increases insulin sensitivity in individuals with metabolic syndrome. Gastroenterology 143(4):913–916, e7

Waldecker M et al (2008) Inhibition of histone-deacetylase activity by short-chain fatty acids and some polyphenol metabolites formed in the colon. J Nutr Biochem 19(9):587–593

Wang Z et al (2011) Gut flora metabolism of phosphatidylcholine promotes cardiovascular disease. Nature 472(7341):57–63

Wang L et al (2013) A small molecule modulates Jumonji histone demethylase activity and selectively inhibits cancer growth. Nat Commun 4:2035

Wang Z et al (2014) Prognostic value of choline and betaine depends on intestinal microbiota-gener-

ated metabolite trimethylamine-N-oxide. Eur Heart J 35(14):904–910
Waterland RA et al (2010) Season of conception in rural gambia affects DNA methylation at putative human metastable epialleles. PLoS Genet 6(12), e1001252
Worthley DL et al (2011) DNA methylation in the rectal mucosa is associated with crypt proliferation and fecal short-chain fatty acids. Dig Dis Sci 56(2):387–396
Wu H, Zhang Y (2014) Reversing DNA methylation: mechanisms, genomics, and biological functions. Cell 156(1–2):45–68
Zempleni J et al (2008) Epigenetic regulation of chromatin structure and gene function by biotin: are biotin requirements being met? Nutr Rev 66(Suppl 1):S46–S48
Zhang H et al (2009) Human gut microbiota in obesity and after gastric bypass. Proc Natl Acad Sci U S A 106(7):2365–2370
Zhang TY et al (2012) Epigenetic mechanisms for the early environmental regulation of hippocampal glucocorticoid receptor gene expression in rodents and humans. Neuropsychopharmacology 38(1):111–123

# The Oral Microbiota

# 4

Nicole B. Arweiler and Lutz Netuschil

**Abstract**

The oral microbiota represents an important part of the human microbiota, and includes several hundred to several thousand diverse species. It is a normal part of the oral cavity and has an important function to protect against colonization of extrinsic bacteria which could affect systemic health. On the other hand, the most common oral diseases caries, gingivitis and periodontitis are based on microorganisms. While (medical) research focused on the planktonic phase of bacteria over the last 100 years, it is nowadays generally known, that oral microorganisms are organised as biofilms. On any non-shedding surfaces of the oral cavity dental plaque starts to form, which meets all criteria for a microbial biofilm and is subject to the so-called succession. When the sensitive ecosystem turns out of balance – either by overload or weak immune system – it becomes a challenge for local or systemic health. Therefore, the most common strategy and the golden standard for the prevention of caries, gingivitis and periodontitis is the mechanical removal of this biofilms from teeth, restorations or dental prosthesis by regular toothbrushing.

**Keywords**

Biofilm • Health-disease-relationship • Periodontitis • Dental plaque

The oral microbiota represents an important part of the human microbiota, and includes, according to different references, several hundred to several thousand diverse species. This diversity comprises several facets:

N.B. Arweiler (✉) • L. Netuschil
Department of Periodontology, University of Marburg, Georg-Voigt-Str., 35039 Marburg, Germany
e-mail: arweiler@med.uni-marburg.de; netuschi@med.uni-marburg.de

1. **Variety**: It is estimated, that a minimum of 700 species occur in the human cavity, from at least 12 phyla (Wade 2013), including even *Archaea*;
2. **Diverse locations**: Saliva; soft tissues like mucosa and the surface(s) of the tongue; hard



A. Schwiertz (ed.), *Microbiota of the Human Body*, Advances in Experimental Medicine and Biology 902, DOI 10.1007/978-3-319-31248-4_4

tissues (teeth) where the dental biofilm (dental plaque) is located in fissures or supra- or subgingival, as well as on hard materials like dentures and, more recently, oral implants;

3. **Intra-oral dislodging**: While quite a lot of the "700 species" prefer specific niches as a habitat, some are found at different locations. For example *Streptococcus mutans* is detected in saliva, in dental (fissure and supragingival) plaque as well as on the tongue. These streptococci, as well as other species, dislodge from one location in the oral cavity to others with a certain mutual relationship.
4. **Age-related microbiological changes**: two different "points of view" are to distinguish: (i) truly age-related changings, and (ii) alterations due to the emergence of teeth, i.e. natural hard surfaces, or to the incorporation of artificial hard surfaces like orthodontic devices, implants, or dentures.
5. **Succession of the oral microbiota – biofilm formation**: This is the change in the composition of the oral microbiota on dental hard surfaces between day 1 (streptococcal, facultative) up to day 7 (Gram-negative rods, spirochetes etc., anaerobes), including the development of sub-systems and so-called "complexes".
6. **Biofilm structure**: The dental plaques on oral hard surfaces represent unique examples of microbial biofilms.
7. "**Health-disease-relationship**" – **significance of oral flora for systemic health**: The normal protective microbiota as compared to (i) caries-related, (ii) periodontitis-related bacteria and (iii) eventually denture stomatitis (Candidiasis) as well as immune stimulation

## 4.1 Variety

About 700 bacterial species inhabit the human mouth, and this quantity seems to be a magic number in quite a lot of articles (Moore and Moore 1994; Aas et al. 2005; Bik et al. 2010; Liu et al. 2012; Jakubovics and Palmer 2013 [Preface]).

About 50 % of these species or phylotypes have not even cultured yet (Marsh 2005; Aas et al. 2005), and in this context Wade (2013) stresses the topic of "uncultivable oral bacteria". This problem touches a serious discussion lasting now for more than a century concerning the viability of (marine and other) bacteria (Winterberg 1898; Ziegler and Halvorson 1935; Postgate 1969; Davey 2011; Netuschil et al. 2014). However, while marine microorganisms seem to be "unculturable" due to dormancy, low temperature or substrate depletion, the problems in culturing oral bacteria are based on their need for very specific nutrients, in part extreme oxygen sensitivity, and, finally, dependence on other neighboring organisms (Wade 2013). For example, some species of the periodontitis-associated microbiota are influenced by the levels of human sexual hormones (Kornman and Loesche 1980, 1981; Jensen et al. 1981). Table 4.1 (from Marsh et al. 2009) lists some properties of the oral microbiota contributing to the difficulty in determining its composition.

More recent genetic analyses disclosed even 10.000 species-level biotypes (Keijser et al. 2008), for example using '454 pyrosequencing' (Voelkerding et al. 2009; Zaura et al. 2009). In spite of the fact that those numbers should be seen with caution, it is a fact that more than tenfold of the species numbers were detected via conventional cultivation (Zaura et al. 2009). As an example Table 4.2 lists only those genera which were found and described since 1990 (from Wade 2013).

## 4.2 Locations of the Oral Microbiota

Figure 4.1 depicts the diverse habitats of the oral microbiota. The **saliva** represents the "planktonic phase" of the oral microbiota. Similar to bacterial laboratory fluid cultures saliva contains up to $10^9$ microorganisms per milliliter, which are swallowed continuously. In this way, about 5 g of bacteria 'disappear' into the stomach daily. Therefore, saliva is not considered to have its own resident microbiota, and that bacterial numbers in saliva, in contrast to dental plaque, do not multiply within the mouth (Marsh et al. 2009). However, saliva is the primary source for the continuous

**Table 4.1** Properties of the oral microflora that contribute to the difficulty in determining its composition (From Marsh et al. 2009)

| Property | Comments |
|---|---|
| High species diversity | The oral microflora, and especially dental plaque, consists of a diverse number of microbial species, some of which are present only in low numbers |
| Surface attachment/coaggregation (coadhesion) | Oral microorganisms attach firmly to surfaces and to each other and, therefore, have to be dispersed without loss of viability |
| Obligate anaerobes | Many oral bacteria lose their viability if exposed to air for prolonged periods |
| Fastidious nutrition/unculturable | Some bacteria are difficult to grow in pure culture and may require specific cofactors etc. for growth |
| | Some groups (e.g. certain spirochaetes; TM7 group) cannot as yet be cultured in the laboratory |
| Slow growth | The slow growth of some organisms makes enumeration time consuming (e.g. they may require 14–21 days incubation) |
| Identification | The classification of many oral microorganisms still remains unresolved or confused; simple criteria for identification are not always available (particularly for some obligate anaerobes) |

**Table 4.2** Recently described bacterial genera with oral representatives (since 1990) (Adapted from Wade (2013))

| Phylum | Genera |
|---|---|
| Actinobacteria | Actinobaculum, Atopobium, Cryptobacterium, Kocuria, Olsenella, Parascardovia, Scardovia, Slackia, Tropheryma |
| Bacteroidetes | Bergeyella, Prevotella, Tannerella |
| Firmicutes | Abiotrophia, Anaerococcus, Aneroglobus, Bulleidia, Catonella, Dialister, Filifactor, Finegoldia, Granulicatella, Johnsonella, Mogibacterium, Parvimonas, Peptoniphilus, Pseudoramibacter, Schwartzia, Shuttleworthia, Solobacterium |
| Proteobacteria | Lautropia, Suttonella |
| Synergistetes | Jonquetella, Pyramidobacter |

bacterial (re)colonization of the diverse oral soft and hard surfaces.

Regarding microbial settlement **shedding surfaces** (mucosal sites) like lips, cheek, palate and tongue have to be differentiated from **non-shedding surfaces**, the natural teeth as well as artificial materials surfaces of fissure sealings, tooth fillings, orthodontic appliances, dentures and also oral implants (Fig. 4.1, Table 4.3; Marsh et al. 2009; Zaura et al. 2009).

Moreover, shedding surfaces, where only monolayers of bacteria originate and which are regularly desquamated (cheek, palate) have to be discriminated from the **tongue** with its 'stable' multilayers of biofilm-like bacteria. It is estimated that the tongue harbours the majority of the microbial burden of the oral cavity, and supports a higher bacterial density and a more diverse microbiota than the other mucosal surfaces; 30 % of the bacterial population detectable by molecular studies were found only on the tongue (Marsh et al. 2009).

On any non-shedding surfaces **dental plaque** starts to form, which meets all criteria for a microbial biofilm (cf. paragraph 6), and is subject to the so-called 'succession' (cf. paragraph 5). Such biofilm formation is found at different locations:

- *Fissure biofilm* (in cavities inside the teeth, approaching the dental pulp) is dominated by facultative species, especially streptococci, causing fissure caries and eventually endodontic problems;
- *Supragingival biofilm* (on the dental enamel adjacent to the gingiva) contains, related to its maturation and thickness, a mixture of facultative and anaerobe species, causing an unspecific gingival inflammation (gingivitis);

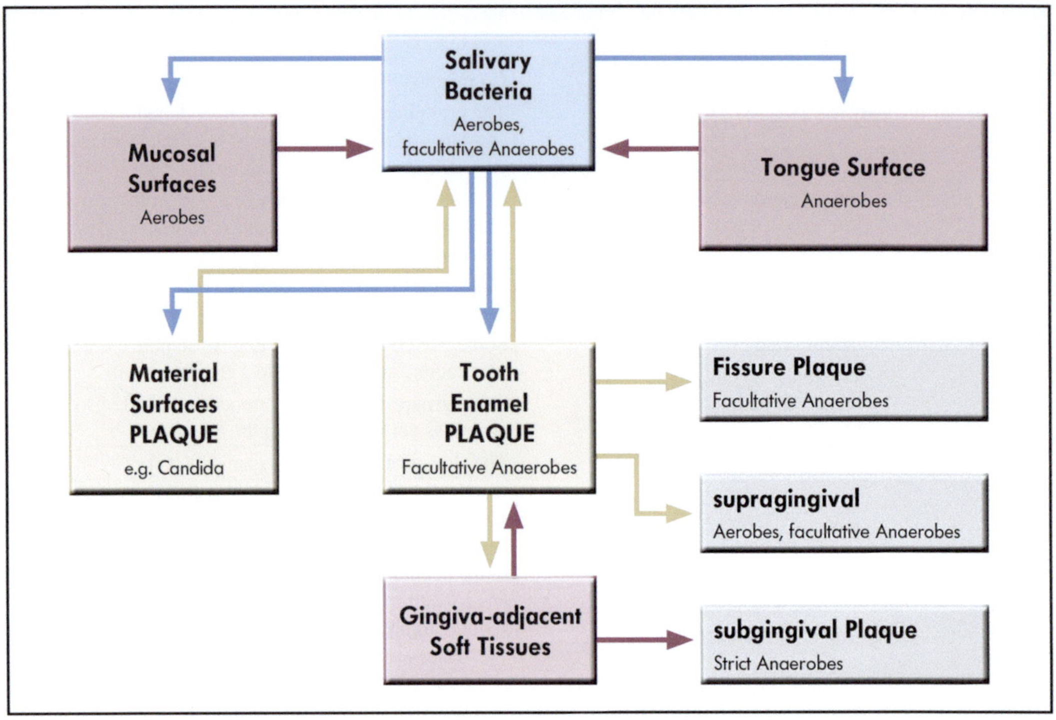


**Fig. 4.1** Colonizing strategy of oral bacteria on the different oral surfaces, and mutual transfers. *Blue*: saliva, planktonic phase; *red*: shedding surfaces; *white*: non-shedding (hard) surfaces

(Fig. 4.2 shows plaque grown on teeth and stained red with disclosure solution)

- Only when supragingival plaque exists for quite a long time harming the gingival crevice, periodontitis may occur due to development of *subgingival plaque*. This type of biofilm contains mainly anaerobe species.
- Plaque on *artificial surfaces* (e.g. dental fillings) resembles mainly the supragingival entity. Denture plaque may harbour *Candida* spp., which may cause 'denture stomatitis'. The microbiota relevant for peri-implant mucositis (analogous to gingivitis) and eventually peri-implantitis (analogous to periodontitis) is not yet well understood (for some further details cf. paragraph 4).

While Table 4.3 describes the diverse habitats of the oral microbiota in general, Table 4.4 (both tables taken from Marsh et al. 2009) summarizes the various bacterial species and groups found at different locations in a normal human oral cavity.

## 4.3 Intra-oral Dislodging: Mutual Transfer

While quite a lot of the "700 species" prefer specific niches as a habitat, some are found at multiple locations. For an example *Streptococcus mutans*, a streptococcal species causing caries, is detected in saliva, in dental (fissure and supragingival) plaque as well as on the tongue (Schlagenhauf et al. 1995) with a certain mutual relationship (Van Houte and Green 1974). A corresponding common semi-quantitative 'Caries Risk Test' is said to rely on the concentration of *S. mutans* in saliva (Beighton 1986). However, the clinician conducting the test is asked to place and turn the test stick on the dorsal surface of the tongue (Schlagenhauf et al. 1995), and therefore the test results reflect rather the amount of *S. mutans* of the tongue biofilm. After suppression of *S. mutans* with chlorhexidine a specific recolonisation pattern could be observed (Emilson et al. 1987). Moreover, it is known that *S. mutans* cannot be eradicated from its oral habitat.

The same holds true regarding periopathogens (mainly Gram-negative anaerobes). Most of these species colonize various niches within the oral cavity, e.g. the oral mucosa, the tongue, the saliva, the periodontal pockets and moreover the oro-pharyngeal area (Quirynen et al. 2001). A specific example represents *Aggregatibacter actinomycetemcomitans*, which is detected not only in supra- and subgingival plaque, but also in saliva and on mucous membranes (Petit et al. 1994). Similar to *S. mutans* these findings have implications for therapeutic measures because the intra-oral translocation of periopathogens may jeopardize the outcome of periodontal therapy (Quirynen et al. 2001).

One study tried to evaluate a sibling relationship on the periodontal conditions. Significant sibship effects were found, among others, for spirochetes on the tongue and in the pockets, for *Porphyromonas gingivalis* on the gingiva and in saliva and for *Prevotella intermedia* in saliva, demonstrating that all those habitats contribute to the overall picture (Van der Velden et al. 1993).

**Table 4.3** Distinct microbial habitats within the mouth (Adapted and complemented from Marsh et al. 2009)

| Habitat | Comments |
|---|---|
| Saliva | Planktonic phase |
| | Continuously swallowed |
| Lips, cheek, palate (shedding, monolayer) | Biomass restricted by desquamation |
| | Some surfaces have specialized host cell types |
| Tongue (shedding, multilayer) | Highly papillated surface |
| | Acts as a reservoir for obligate anaerobes |
| Natural (teeth) and artificial hard surfaces (dental materials) (non-shedding, multilayers) (dental plaque biofilms) | Non-shedding surfaces enabling large masses of microbes to accumulate |
| | Teeth have distinct surfaces for microbial colonization; each surface (e.g. fissures, smooth surfaces, approximal, gingival crevice) will support a distinct microflora because of their intrinsic biological properties |

## 4.4 Inter-oral Transmission of Bacteria and Age-Related Microbiological Changes

Teo different "points of view" are associated with age-related changes: (i) truly age-related changes, and (ii) alterations due to the emergence of teeth,

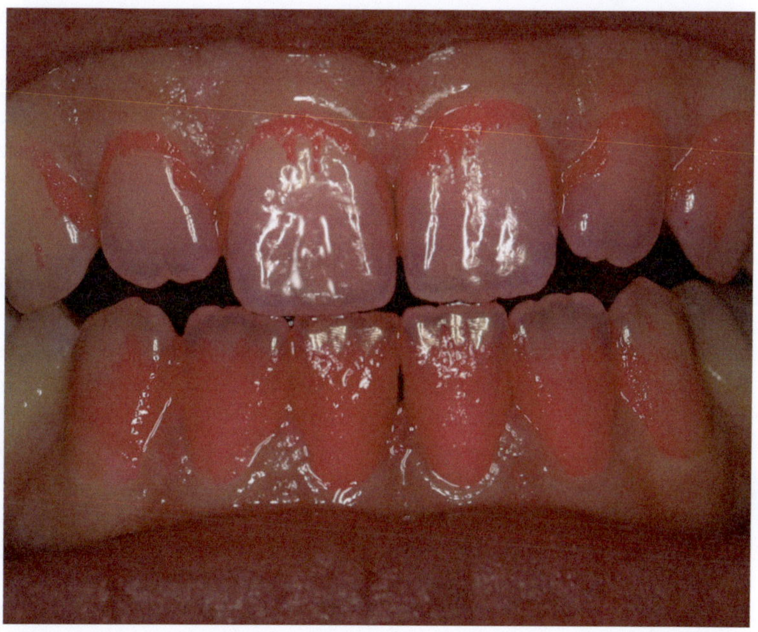

**Fig. 4.2** Dental biofilm on teeth visualized by dental disclosure solutions

**Table 4.4** Relative proportions of some cultivable bacterial populations at different sites in the normal oral cavity (From Marsh et al. 2009)

| Bacterium | Saliva | Buccal mucosa | Tongue dorsum | Supragingival plaque |
|---|---|---|---|---|
| *Streptococcus sanguinis* | 1 | 6 | 1 | 7 |
| *S. salivarium* | 3 | 3 | 6 | 2 |
| *S. oralis/S. mitis* | 21 | 29 | 33 | 23 |
| Mutans streptococci | 4 | 3 | 3 | 5 |
| *Actinomyces naeslundii* | 2 | 1 | 5 | 5 |
| *A. odontolyticus* | 2 | 1 | 7 | 13 |
| *Haemophilus* spp | 4 | 7 | 15 | 7 |
| *Capnocytophaga* spp | <1 | <1 | 1 | <1 |
| *Fusobacterium* spp | 1 | <1 | <1 | <1 |
| *Black-pigmented anaerobes* | <1 | <1 | 1 | +[a] |

[a]Detected on occasions

i.e. natural hard surfaces, or to the implementation of artificial hard surfaces like orthodontic devices, implants, or dentures. At first glance the microbiota does not change over decades in age-cohorts between 20 and <70 years. Percival et al. (1991) detected no age-related changes in caries associated mutans streptococci and in periodontitis-related spirochetes. However, differences were found in age groups >70 years differences were found regarding lactobacilli, staphylococci, and yeasts, which increased significantly. These increases were not related to denture-wearing or disease. It has been assumed that age-dependent multi-morbidity and multifold usage of medicaments, which are well known to diminish saliva production, and the subsequent acidification of the oral milieu are responsible for this change (Kraneveld et al. 2012). However, these studies did not consider children and adolescents. Moreover, the caries-related and the periodontitis-associated microbiota have to be distinguished.

Concerning caries alterations due to point (ii) are of importance. An obvious example represents the emergence of teeth occurring during the 6th-7th month on. Only then streptococci like *S. mutans* and *S. sobrinus* are able to colonize the hard enamel surfaces. Surprisingly, it is not until 1 year later that a corresponding "window of infectivity" can be found and designated (Caufield et al. 1993) between 19 and 31 month of age, with a median of 26 month.

In later years, when orthodontic appliances are incorporated (e.g. in Germany, 45 % of 12 years old and 58 % of 15 years old wear such appliances, Micheelis et al. 2006) such new material surfaces contribute to a tremendous increase in the salivary levels of caries-related bacteria like *S. mutans*. Last not least when dentures are used, *Candida* spp. may increase (Gendreau and Loewy 2011; Salerno et al. 2011). The corresponding literature reflects an ambiguous picture. On the one hand it is described that *Candida* overgrowth is not associated with denture-wearing (Percival et al. 1991; Kraneveld et al. 2012), while on the other hand "denture stomatitis" has become a standard term regarding this aspect in dentistry of the elderly.

It is also well established that transmission of anaerobic bacteria, in part periopathogens, occurs during the first year in children (Könönen 1999), where *Veillonella* spp. and *P. melaninogenica* were found even after 2 months, while *A. actinomycetemcomitans* seems to be a 'late colonizer' emerging between the ages of 4–7 years (Alaluusua and Asikainen 1988; Könönen 1999). For a short overview see Table 4.5.

Similar to *S. mutans*, these species are mostly transferred via the maternal saliva (Könönen et al. 1992); in one specific case even the transmittance of *A. actinomycetemcomitans* from a dog to a child was reported (Preus and Olsen 1988). The pattern of colonizing of these anaerobes is even influenced by the emergence of the primary dentition (Könönen et al. 1994) in spite of the fact that these species have no impact on diseases of intraoral hard tissues like enamel. In 2- to12- year-old children periopathogens are frequently detected (Okada et al. 2001).

**Table 4.5** The effect of tooth eruption on the composition of the cultivatable oral microflora in young children (From Marsh et al. 2009)

| | Percentage isolation frequency | |
|---|---|---|
| | At mean age | |
| Bacterium | 3 months | 32 months |
| *Prevotella melaniogenica* | 76 | 100 |
| Non-pigmented *Prevotella* | 62 | 100 |
| *Prevotella loescheii* | 14 | 90 |
| *Prevotella intermedia* | 10 | 67 |
| *Prevotella denticola* | Not detected | 71 |
| *Fusobacterium nucleatum* | 67 | 100 |
| *Fusobacterium* spp. | Not detected | 71 |
| *Selenomonas* spp. | Not detected | 43 |
| *Capnocytophaga* spp. | 19 | 100 |
| *Leptotrichia* spp. | 24 | 71 |
| *Campylobacter* spp. | 5 | 43 |
| *Eikenella corrodens* | 5 | 57 |
| *Veillonella* spp. | 63 | 63 |

Intrafamilial transmission has been shown (Alaluusua et al. 1991; Petit et al. 1994), especially between spouses, where disease-related bacteria are transferred between (clinically) healthy and periodontally diseased family members (Offenbacher et al. 1985; Saarela et al. 1993). It was evident that horizontal transmission between spouses ranged between 14 % and 60 % for *A. actinomycetemcomitans*, and between 30 % and 75 % for *P. gingivalis*, while vertical transmission was estimated between 30 % and 60 % for *A. actinomycetemcomitans*, but only rare vertical transmission of *P. gingivalis* was observed (Van Winkelhoff and Boutaga 2005). Thus, it is a matter of discussion whether this phenomenon hampers the treatment outcome (Von Troil-Linden et al. 1997; Kleinfelder et al. 1999).

After early childhood there seems no further change till puberty (Könönen 1999), while during puberty alterations are found regarding *Veillonella* spp., *Prevotella denticola* and *Prevotella melaninogenica* (Gusberti et al. 1990; Moore et al. 1993). Because, as previously mentioned, some members of the periodontitis-associated microbiota are influenced by the levels of human hormones (Kornman and Loesche 1980; Jensen et al. 1981), Moore et al. (1993) looked for an association with testosterone levels. However, no such relationship could be established.

A specific and in part still neglected problem is the peri-implant mucositis and peri-implantitis (which is called "the periodontitis of the implant"). Implant surfaces are as easily and as rapidly colonized by bacteria as teeth (Fürst et al. 2007), however, different colonization patterns seem to exist on implants as compared to teeth (Salvi et al. 2008). In general, only few investigations exist assessing the microbiota of implants when compared to the bulk of literature regarding periodontitis – contributing to a confusing picture. The peri-implant microbiota is described to be quite similar to that of periodontitis (Leonhardt et al.1993), but with some relevant differences (Persson and Renvert 2013). For example *Staphylococcus aureus* is more common in peri-implant plaque (Rams et al. 1990), because this germ is attracted by titanium surfaces (Harris and Richards 2004). Figure 4.3 shows a SEM picture of a biofilm on an implant which had to be explanted due to periimplantitis.

These local bacterial changes are also a systemic challenge and stimulation for the immune system. This aspect will be discussed in Sect. 4.7.

## 4.5 Succession of the Oral Microbiota: Biofilm Formation

Because **Saliva** is frequently swallowed, no succession can occur. Salivary bacteria regularly colonize the **mucosal cells** (Slots and Gibbons 1978). In spite of the fact that the soft tissues represent 80 % of the surfaces prone to bacterial colonization (Collins and Dawes 1987) no pathogenetic problems arise thereof, because the mucosal cells desquamate and are swallowed, similar to saliva. Therefore, only a non-pathogen monolayer consists on mucosal surfaces.

In contrast hard tissues are immediately covered by the '**pellicle**' (Sönju 1987; Hannig 1997) (for example after meticulous tooth cleaning measures) and concomitantly colonized by bacteria (Rönström et al. 1977; Kolenbrander and London 1993). Following this first phase the **plaque biofilm** microbiota changes steadily

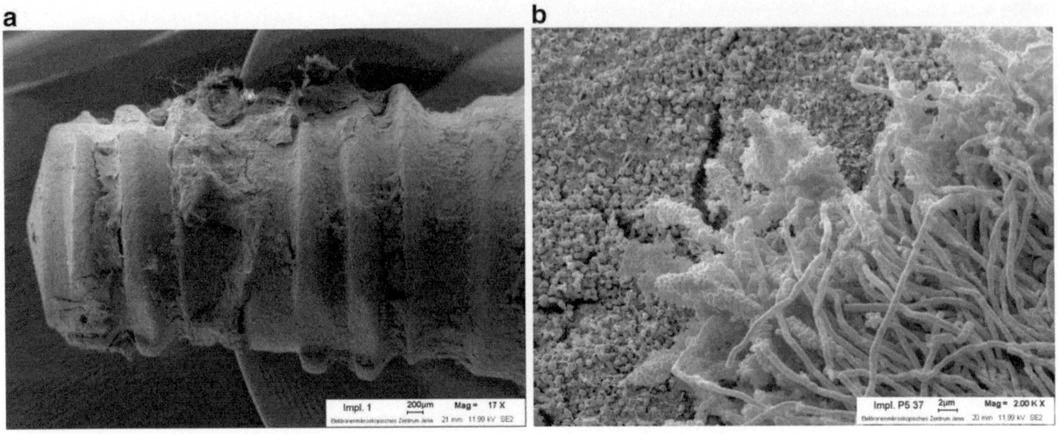


**Fig. 4.3** (**a**, **b**): SEM picture of a (highly diverse) biofilm on an implant which had to be explanted due to periimplantitis (Thanks also to Prof. S. Nietzsche, University of Jena)

from day to day, a process called **succession** (Theilade et al. 1966; Ritz 1967; Marsh 1990). An immediate streptococci-dominated 'primary flora' changes during 7 days of development to an anaerobic 'climax community', characterized by Gram-negative rods (Morhardt and Fitzgerald 1980). Due to different localizations as well as diverse exogenous influencing factors, plaque of different thickness and bacterial composition develops, not only on a macroscopic scale but also at the micro-ecological level, as related to $O_2$-tension, local pH, matrix structure and availability of nutritive substances (Marsh et al. 2009). Thus diverse sub-systems develop (Garlichs et al. 1974; Morhardt and Fitzgerald 1980; van Palenstein Helderman 1981; Babaahmadi et al. 1998). One example is the site-specific distribution pattern of periodontitis-relevant micro-organisms as described by Tanner et al. (1998).

## 4.6 Phases of Biofilm Development

Several different phases are characterized during succession. This holds true for dental but also for all other natural occurring biofilms – such as medical or environmental biofilm. Dental biofilm formation can be visualized for example by confocal laser scanning microscopy (CLSM) on enamel slabs, presented in Fig. 4.4a–d.

**Induction**: This first phase is characterized by the formation of the aforementioned pellicle as a 'conditioning film' or 'linking film' (Busscher and van der Mei 1997), but the first bacteria are also sometimes already visible (Marsh and Bradshaw 1995; Hannig 1999) (see Fig. 4.4a).

**Accumulation**: This second step includes different topics like bacterial adhesion, bacterial growth ("planar colonisation") and so called 'quorum sensing' (cf. Chap. 6).

**Existence**: The third phase, when occurring, is characterized by equilibrium between growth and concomitant decomposition via so called "biofilm erosion" and "biofilm sloughing", whereby cells and cell clusters are teared off for settling on other surfaces.

The event of a succession in the oral microbiota was confirmed for subgingival plaque by Socransky et al. (1998) and Haffajee et al. (1999). Using molecular biology techniques the authors assessed and grouped diverse bacterial "inhabitants" of about 13,000 plaque samples in so called 'complexes'. One 'yellow complex' comprised mainly of streptococci, which in accordance to other authors (Theilade et al. 1966; Kolenbrander et al. 1999) were shown to represent the earliest colonizers. In a second 'orange complex' different species were grouped, the most important being *Fusobacterium nucleatum*. This species has a high ability to coaggregate with other bacteria

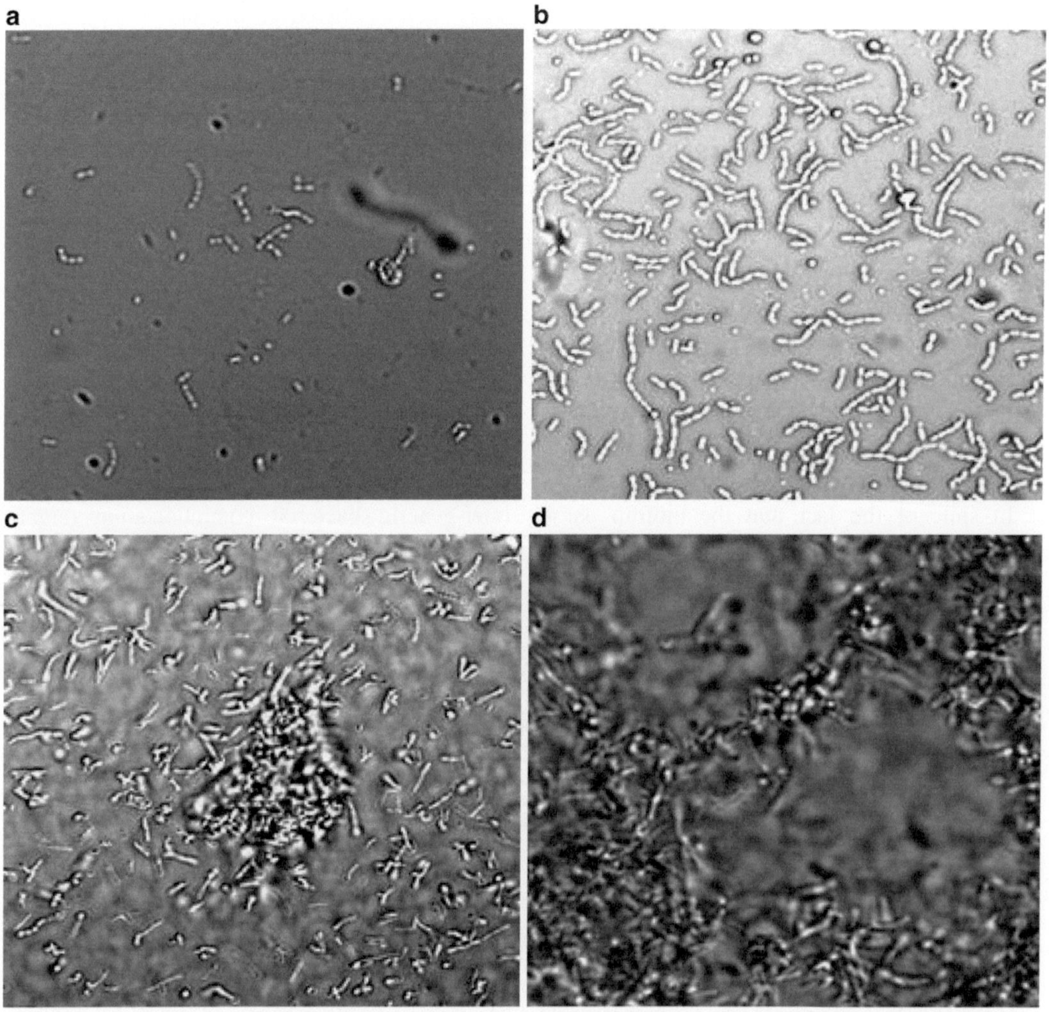

**Fig. 4.4** (a–d): Biofilm formation on dental enamel tracked by confocal laser scanning microscopy (gray modus)

(Kolenbrander et al. 1989), thus "bridging" species of the earlier 'yellow' with those of the late 'red complex'. This last complex comprising of *P. gingivalis*, *Tannerella forsythus* and *Treponema denticola* was strongly associated with the maintenance and worsening of periodontitis. Interestingly, *F. nucleatum* proved to be the most frequent anaerobe species in infants' mouths at 1 year of age (Könönen 1999), a finding perfectly fitting with the concept of the crucial role of *F. nucleatum* in intergeneric coaggregation and biofilm formation (Kolenbrander et al. 1989). In sum a succession from mainly facultative species (in part involved in dental caries) to a more and more anaerobe community occurs, which is finally responsible for the etiology of gingivitis and periodontitis (Theilade et al. 1966; van Palenstein Helderman 1981; Socransky et al. 1998; Haffajee et al. 1999).

## 4.7 Dental Plaque: A Typical Biofilm

There are several definitions of the term biofilm, for example "*matrix enclosed bacterial populations adherent to each other and/or to surfaces or interfaces*" (Costerton et al. 1995; Costerton and Lewandowsky 1997); "*a biofilm will form on any surface that is exposed to microbes, water, and a small amount of nutrient*" (Wimpenny 1997); or, somewhat more detailed "*A biofilm is defined as bacterial aggregates, usually existing as closely associated communities, that adhere to assorted natural or artificial surfaces, usually in aqueous environment that contains a sufficient concentration of nutrients to sustain the metabolic needs of the microbiota*" (Listgarten 1999). In this respect dental plaque shows the general characteristics of biofilms (Wimpenny et al. 2000; Costerton et al. 1999); moreover, as mentioned, it harbours a plethora of bacterial species, and thus, is extremely heterogeneous (Costerton et al. 1999; ten Cate 2006; Paster et al. 2006; Zaura et al. 2009). For a short overview see Table 4.6.

In spite of the fact that the biofilm phase worldwide comprises 80–90 % of microorganisms, the (medical) research focused on the planktonic phase of bacteria over the last 100 years. Biofilms contain 1000-fold more bacteria per gram than the planktonic phase; on the other hand they are by factor 500–1000 more resistant against antibacterial compounds. It is noteworthy, that both of these factors are equal in their importance. However, the higher density is not the main reason for the increased resistance. A few general examples are on the enormous importance of biofilms in ecology, medicine and industry are presented in Table 4.6.

Donlan and Costerton (2002) state that not only the architecture and other readily observable characteristics like adherence and extracellular matrices, are important for differences in comparison to planktonic cells, but also inherent attributes such as altered growth rate and antibiotic resistance: biofilm organisms transcribe genes that planktonic cells do not. Moreover, bacteria bound in the environment of biofilms exert a substantial resistance against detergents, antibiotics or other antibacterial compounds, and phagocytosis due to "persisters" (Donlan and Costerton 2002; Obst et al. 2006; Anwar et al. 1992). The underlying mechanisms are complex and multifactorial (del Pozo and Patel 2007). A modern definition thus needs to contain the facets that biofilm organisms exhibit an altered phenotype with respect to growth rate, gene transcription and 'quorum sensing'. This latter phenomenon, also called 'biofilm signaling', describes the inter-generic bacterial communication (signal transduction), which depends on cell density and occurs in maturating biofilms (Kaiser and Losick 1993; Fuqua et al.1996; Costerton et al. 1999; Prosser 1999).

Concerning **methods in biofilm research** there are some crucial prerequisites when evaluating oral (plaque) biofilms: (1) Intraoral splint systems, which enable the undisturbed accumulation of dental biofilms on the surface(s) of native enamel slabs (Auschill et al. 2004; Arweiler et al. 2004) or dental materials (Auschill et al. 2002); including (2) the concomitant formation of a native pellicle (Hannig 1997, 1999).

Traditionally, the oral tooth-related microbiota was and still is assessed either by conventional microbiological methods (cultivation; Theilade et al. 1966; Theilade and Theilade 1970; Mikkelsen 1993) or by electron microscopy

**Table 4.6** Characteristics of biofilms

| Parameter | Planktonic phase | Biofilm |
|---|---|---|
| General definition | Free floating, non-adherend, not localized | Adherend, localized |
| Density | $10^8$–$10^9$ Bacteria/mL = $10^8$–$10^9$ Bacteria/gram | $10^{11}$–$10^{12}$ Bacteria/cm$^3$ = $10^{11}$–$10^{12}$ Bacteria/gram |
| Occurence | 10–20 % | 80–90 % |
| Research focus | 1880–1980 | 1960–to date |
| Resistance against antibac+terial compounds | Generally low | Generally high or very high |
| (Medical) importance | E.g. Legionella; E. coli | Med. catheters, lenses |
| | | Oral diseases |
| | | Food/paper/oil industries |
| | | Navigation |

(TEM and SEM; Listgarten 1965; Theilade and Theilade 1970; Saxton 1973; for review cf. Newman and Wilson 1999). Furthermore, vital (fluorescence) staining techniques were used to elucidate the portion of vital or dead bacteria in the dental biofilm (Netuschil et al. 1989, 2014), which can also visualize the effect of antibacterial agents by CLSM (Fig. 4.5, vitalfluorescence (VF); Fig. 4.6, VF combined with CLSM). More recently the FISH technology (Fluorescence in situ hybridization) was introduced to plot specific bacterial species and to depict the distribution of them in a biofilm network (Al-Ahmad et al. 2007; Fig. 4.7). Thus, different "visualizing" methods were combined with CLSM to reveal the three-dimensional architecture of oral biofilms (Netuschil et al. 1998, 2014; Auschill et al. 2002, 2004; Arweiler et al. 2004, 2013; Zaura et al. 2001; Al-Ahmad et al. 2007, 2009, 2010).

## 4.8 "Health-Disease-Relationship" and Significance of Oral Flora for Systemic Health

As mentioned, highly diverse oral microbiota is a normal part of the oral cavity. It has an important function to protect against colonization of extrinsic bacteria which could affect systemic health.

Former recommendations aimed on a high standard of oral hygiene (of parents) to prevent microbial colonization of the infantine oral cavity. Newer findings show that parental sucking of pacifiers or spoons leads to a risk reduction of allergy development possibly via immune stimulation by the transferred microbes (Hesselmar et al. 2013).

On the other hand, the most common oral diseases caries, gingivitis and periodontitis are based on microorganisms. Bacteria are a necessary but not sufficient requirement for the development of these diseases. It is generally assumed that ecological conditions (especially of the host) play a key role in the development of these diseases. The same hold true for infection with candida species and the formation of a denture stomatitis (Candidiasis). Most people are carriers of candida but oral candidiasis is very rare.

The role of periodontal disease as a risk factor in the development and/or progression of systemic diseases such as diabetes, rheumatoid arthritis, cardiovascular disease, adverse pregnancy outcomes, head-and-neck cancer) has been subject of many investigations in recent years (Han et al. 2014).

## 4.9 Prospects

While vaccination is due to the high diversity of the oral biofilm not an alternative, brand new concepts aim to strengthen the normal, protective microbiota. Probiotics are known as promoter of a natural microbiota, e.g. following the administration of antibiotics. Meanwhile, numerous probiotic products containing bacteria such as lactobacilli or *E. faecalis* are commercially available and have been widely-used for their health benefits. From a dental point of view, colonization of the oral cavity and particularly the oral biofilm with probiotic bacteria is feared, since they are able to ferment sugars. This results in acid production, which causes dissolution of the enamel and caries development. This factor and also the question if probiotic bacteria can reside and persist in the oral biofilm, have to be answered in future research. Most recent research provide an indication that probiotic bacteria such as lactobacilli will not prevail within the oral biofilm but can significantly suppress S. mutans spp. (Al Ahmad et al. 2014).

## 4.10 Summary

The microbiota is an important part of our oral cavity. However, when this sensitive ecosystem turns out of balance – either by overload or weak immune system – it becomes a challenge for local or systemic health. Therefore, the most common strategy and the golden standard for the prevention of caries, gingivitis and periodontitis is the mechanical removal of this biofilms from teeth, restorations or dental prosthesis by regular tooth-

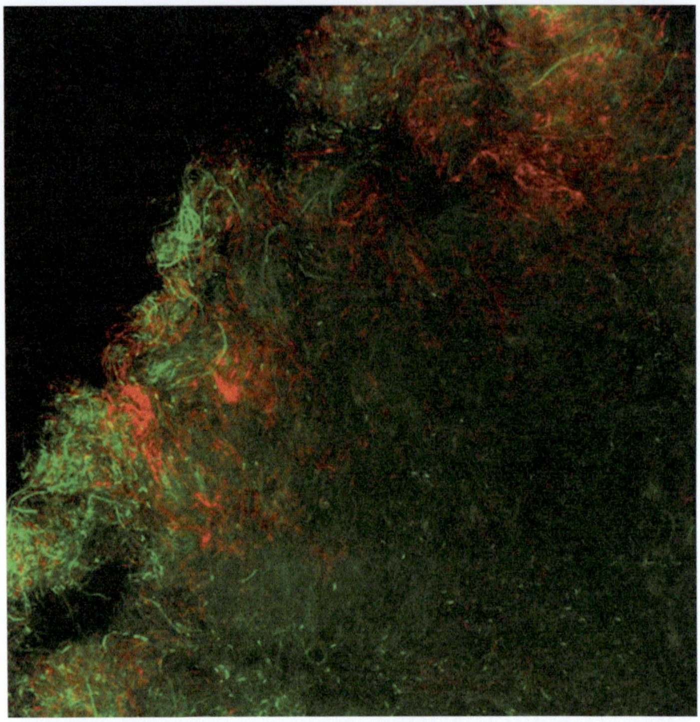

**Fig. 4.5** Vital fluorescence staining to elucidate the portion of vital (*green*) and dead (*red*) bacteria in dental biofilm

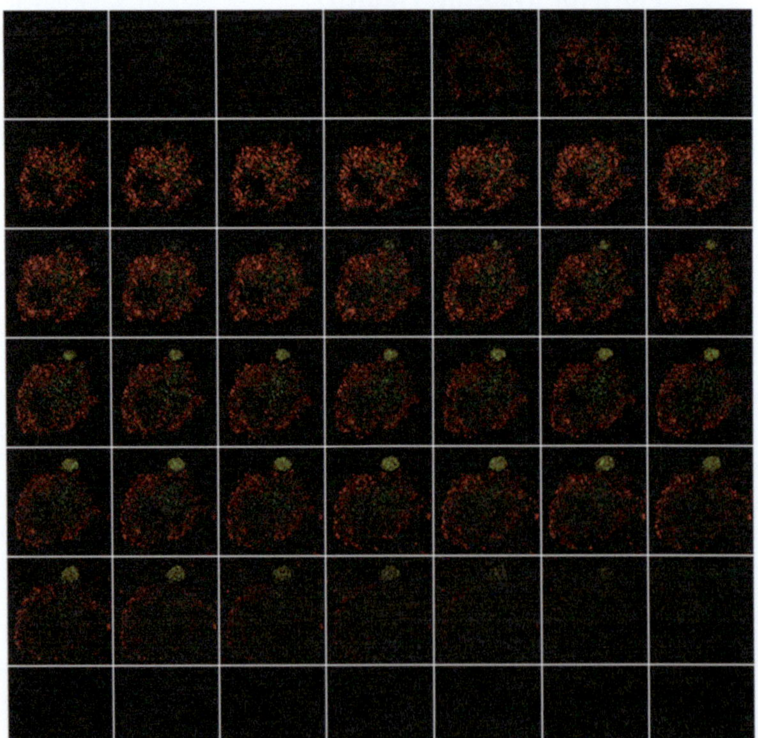

**Fig. 4.6** Three dimensional architecture of oral biofilm and distribution of vital and dead bacteria after regular rinsing with chlorhexidine (antiseptic agent). Still a "vital core" is *left* which shows the highly resistant nature of biofilms

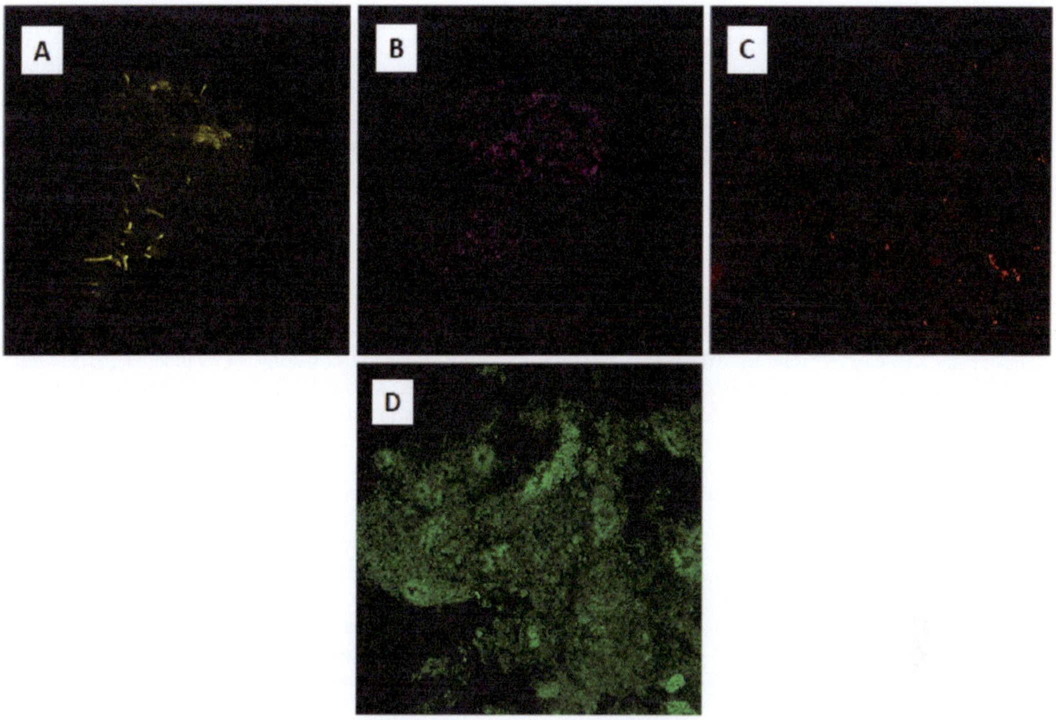

**Fig. 4.7** Confocal micrographs of 3-day-old dental plaque biofilm hybridized with four different specific probes (Multiplex-FISH) as described in Al-Ahmad et al. (2007). (**A**) *yellow*, *F. nucleatum*-specific probe; (**B**) *magenta*, *Streptococcus spp.*-specific probe; (**C**) *red*, *Veillonella spp.*-specific probe; (**D**) *Green*, eubacteria-specific Probe – all bacteria. The distribution of specific bacteria in the biofilm network can be determined by comparison with eubacteria using a software program. (By courtesy of Prof. Dr. Ali Al-Ahmad)

brushing, daily interdental cleaning and by dental professionals on a regular basis. This has to be trained and adapted from childhood. It also guarantees that biofilms which start to form within a few minutes after removal only develop into the phase of an early, thin and not anaerobe biofilm.

## References

Aas JA, Paster BJ, Stokel LN, Olsen I, Dewhirst FE (2005) Defining the normal bacterial flora of the oral cavity. J Clin Microbiol 43:5721–5732

Al-Ahmad A, Wunder A, Auschill TM, Follo M, Braun G, Hellwig E, Arweiler NB (2007) The in vivo dynamics of Streptococcus spp., Actinomyces naeslundii, Fusobacterium nucleatum and Veillonella spp. in dental plaque biofilm as analysed by five-colour multiplex fluorescence in situ hybridization. J Med Microbiol 56:681–687

Al-Ahmad A, Follo M, Selzer AC, Hellwig E, Hannig M, Hannig C (2009) Bacterial colonization of enamel in situ investigated using fluorescence in situ hybridization. J Med Microbiol 58:1359–1366

Al-Ahmad A, Wiedmann-Al-Ahmad M, Faust J, Bächle M, Follo M, Wolkewitz M, Hannig C, Hellwig E, Carvalho C, Kohal R (2010) Biofilm formation and composition on different implant materials in vivo. J Biomed Mater Res B Appl Biomater 95:101–109

Al-Ahmad A, Hellwig E, Follo M, Auschill TM, Arweiler NB (2014) Probiotic lactobacilli do not integrate into oral biofilm in situ. In: The ESCMID Study Group for Biofilms meeting, 09–10 October 2014, Rome

Alaluusua S, Asikainen S (1988) Detection and distribution of Actinobacillus actinomycetemcomitans in the primary dentition. J Periodontol 59:504–507

Alaluusua S, Asikainen S, Lai C-H (1991) Intrafamilial transmission of Actinobacillus actinomycetemcomitans. J Periodontol 62:207–210

Anwar H, Strap JL, Costerton JW (1992) Establishment of aging biofilms: possible mechanism of bacterial resistance to antimicrobial therapy. Antimicrob Agents Chemother 36:1347–1351

Arweiler NB, Hellwig E, Sculean A, Hein N, Auschill TM (2004) Individual vitality pattern of in situ dental biofilms at different locations in the oral cavity. Caries Res 38:442–447

Arweiler NB, Netuschil L, Beier D, Grunert S, Heumann C, Altenburger MJ, Sculean A, Nagy K, Al-Ahmad A, Auschill TM (2013) Action of food preservatives on 14-days dental biofilm formation, biofilm vitality, and biofilm-derived enamel demineralisation in situ. Clin Oral Invest. [Epub ahead of print] doi:10.1007/s00784-013-1053-9

Auschill TM, Arweiler NB, Brecx M, Reich E, Sculean A, Netuschil L (2002) The effect of dental restorative materials on dental biofilm. Eur J Oral Sci 110:48–53

Auschill TM, Hellwig E, Sculean A, Hein N, Arweiler NB (2004) Impact of the intraoral location on the rate of biofilm growth. Clin Oral Invest 8:97–101

Babaahmady KG, Challacombe SJ, Marsh PD, Newman HN (1998) Ecological study of *Streptococcus mutans*, *Streptococcus sobrinus* and *Lactobacillus spp*. at subsites from approximal dental plaque from children. Caries Res 32:51–58

Beighton D (1986) A simplified procedure for estimating the level of Streptococcus mutans in the mouth. Br Dent J 160:329–330

Bik EM, Armitage GC, Loomer P, Emerson J, Mongodin EF, Nelson KE, Gill SR, Raser-Liggett CM, Relman DA (2010) Bacerial diversity in the oral cavity of 10 healthy individuals. ISME J 4:962–974

Busscher HJ, van der Mei HC (1997) Physico-chemical interactions in initial microbial adhesion and relevance for biofilm formation. Adv Dent Res 11:24–32

Caufield PW, Cutter GR, Dasanayake AP (1993) Initial Acquisition of mutans streptococci by infants: evidence for a discrete window of infectivity. J Dent Res 72:37–45

Collins LMC, Dawes C (1987) The surface area of the adult human mouth and thickness of the salivary film covering the teeth and oral mucosa. J Dent Res 66:1300–1302

Costerton JW, Lewandowski Z (1997) The biofilm lifestyle. Adv Dent Res 11:192–195

Costerton JW, Lewandowski Z, Caldwell DE, Korber DR, Lappin-Scott HM (1995) Microbial biofilms. Ann Rev Microbiol 49:711–745

Costerton JW, Cook G, Lamont R (1999) The community architecture of biofilms: dynamic structures and mechanisms. In: Newman HN, Wilson M (eds) Dental plaque revisited. BioLine, Antony Rowe Ltd, Chippenham, pp 5–14

Davey HM (2011) Life, death, and in-between: meanings and methods in microbiology. Appl Environ Microbiol 77:5571–5576

del Pozo JL, Patel R (2007) The challenge of treating biofilm-associated bacterial infections. Clin Pharmacol Ther 82:204–209

Donlan RM, Costerton JW (2002) Biofilms: survival mechanisms of clinically relevant microorganisms. Clin Microbiol Rev 15:167–193

Emilson CG, Lindquist B, Wennerholm K (1987) Recolonization of human tooth surfaces by Streptococcus mutans after suppression by Chlorhexidin treatment. J Dent Res 66:150–1508

Fuqua C, Winans SC, Greenberg EB (1996) Census and consensus in bacterial ecosystems: the LuxR-LuxI family of quorum-sensing transcriptional regulators. Ann Rev Microbiol 50:727–751

Fürst MM, Salvi GE, Lang NP, Persson GR (2007) Bacterial colonization immediately after installation on oral titanium implants. Clin Oral Implants Res 18:501–508

Garlichs UA, Brandau H, Bössmann K (1974) Histotopochemical determination of metabolic activity of carbohydrate metabolism in plaque from sound and carious enamel. Caries Res 8:234–248

Gendreau L, Loewy ZG (2011) Epidemiology and etiology of denture stomatitis. J Prosthodont 20:251–260

Gusberti FA, Mombelli A, Lang NP, Minder CE (1990) Changes in subgingival microbiota during puberty: a 4-year longitudinal study. J Clin Periodontol 17:685–692

Haffajee AD, Socransky SS, Feres M, Ximenez-Fyvie LA (1999) Plaque microbiology in health and disease. In: Newman HN, Wilson M (eds) Dental plaque revisited. BioLine, Antony Rowe Ltd, Chippenham, pp 255–282

Han YW, Houcken W, Loos BG, Schenkein HA, Tezal M (2014) Periodontal disease, atherosclerosis, adverse pregnancy outcomes, and head-and-neck cancer. Adv Dent Res 26:47–55

Hannig M (1997) Transmission electron microscopic study of in vivo pellicle formation on dental restorative materials. Eur J Oral Sci 105:422–433

Hannig M (1999) Ultrastructural investigation of pellicle morphogenesis at two different intraoral sites during a 24-h period. Clin Oral Invest 3:88–95

Harris LG, Richards RG (2004) Staphylococcus aureus adhesion to different treated titanium surfaces. J Mater Sci Mater Med 15:311–314

Hesselmar B, Sjöberg F, Saalman R, Aberg N, Adlerberth I, Wold AE (2013) Pacifier cleaning practices and risk of allergy development. Pediatrics 131:1829–1837

Jakubovics NS, Palmer RJ Jr (2013) Oral microbial ecology – current research and new perspectives. Caister Academic Press, Norfolk

Jensen J, Liljemark W, Bloomquist C (1981) The effect of female sex hormones on subgingival plaque. J Periodontol 52:599–602

Kaiser D, Losick R (1993) How and why bacteria talk to each other. Cell 73:873–885

Keijser BJ, Zaura E, Huse SM, van der Vossen JM, Schuren FH, Mntijn RC, ten Cate JM, Crielaard W (2008) Pyrosequencing analysis of the oral microflora of healthy adults. J Dent Res 87:1016–1020

Kleinfelder JW, Müller RF, Lange DE (1999) Intraoral persistence of Actinobacillus actinomycetemcomitans in periodontally healthy subjects following treatment of diseases family members. J Clin Periodontol 26:583–589

Kolenbrander PE, London J (1993) Adhere today, here tomorrow: oral bacterial adherence. J Bacteriol 175:3247–3252

Kolenbrander PE, Andersen RN, Moore LHV (1989) Coaggregation of Fusobacterium nucleatum, Selenomonas flueggei, Selenomonas infelix, Selenomonas noxia, and Selenomonas sputigena with strains from 11 genera of oral bacteria. Infect Immun 57:3194–3203

Kolenbrander PE, Andersen RN, Clemans DL, Whittaker CJ, Klier CM (1999) Potential role of functionally similar coaggregation mediators in bacterial succession. In: Newman HN, Wilson M (eds) Dental plaque revisited. BioLine, Antony Rowe Ltd, Chippenham, pp 171–186

Könönen E (1999) Oral colonization by anaerobic bacteria during childhood: role in health and disease. Oral Dis 5:278–285

Könönen E, Jousimies-Somer H, Asikainen S (1992) Relationship between oral gramnegative anaerobic bacteria in saliva of the mother and the colonization of her edentulous infant. Oral Microbiol Immunol 7:273–276

Könönen E, Asikainen S, Saarela M, Karjalainen J, Jousimies-Somer H (1994) The oral gram-negative anaerobic microflora in young children: longitudinal changes from edentulous to dentate mouth. Oral Microbiol Immunol 9:136–141

Kornman KS, Loesche WJ (1980) The supragingival microbial flora during pregnancy. J Periodontal Res 15:111–112

Kornman KS, Loesche WJ (1981) Effects of estradiol and progesterone on Bacteroides melaninogenicus and Bacteroides gingivalis. Infect Immun 35:256–257

Kraneveld EA, Buijs MJ, Bonder MJ, Visser M, Keijser BJ, Crielaard W, Zaura E (2012) The relation between oral candida load and bacterial microbiome profiles in Dutch older adults. PLoS One 8(8):e42770. doi:10.1371/journal.pone.0042770

Leonhardt A, Adolfsson B, Lekholm U, Wikström M, Dahlén E (1993) A longitudinal microbiological study on osseointegrated titanium implants in partially edentulous patients. Clin Oral Implants Res 4:113–120

Listgarten MA (1965) Electron microscopic observation on the bacterial flora of acute necrotizing ulcerative gingivitis. J Periodontol 36:328–339

Listgarten MA (1999) Formation of dental plaque and other oral biofilms. In: Newman HN, Wilson M (eds) Dental plaque revisited. BioLine, Antony Rowe Ltd, Chippenham, pp 187–210

Liu B, Falle LL, Klitgord N, Mazumdar V, Ghodsi M, Sommer DD, Gibbons TH, Teangen TJ, Chang Y-C, Li S, Stine OC, Hasturk H, Kasif S, Segrè D, Pop M, Amar S (2012) Deep sequencing of the oral microbiome reveals signatures of periodontal disease. PLoS One 7:e37919. doi:10.1371/journal.pone.0037919

Marsh PD (1990) Microbial succession in relation to enamel demineralisation. Microb Ecol Health Dis 3:i–iii

Marsh PD (2005) Dental plaque: biological significance of a biofilm and community life-style. J Clin Periodontol 32(Suppl 6):7–15

Marsh PD, Bradshaw DJ (1995) Dental plaque as a biofilm. J Ind Microbiol 15:169–175

Marsh PD, Martin MV, Lewis MAO, Williams DW (2009, 2010, 2013) Oral microbiology, 5th ed. Churchill Livingstone Elsevier, Edinburgh

Micheelis W, Schiffner U, Hoffmann T, Kerschbaum T, John MT (2006) Vierte Deutsche Mundgesundheitsstudie (DMS IV). Deutscher Zahnärzte Verlag, Köln

Mikkelsen L (1993) Influence of sucrose intake on saliva and number of microorganisms and acidogenic potential in early dental plaque. Microb Ecol Health Dis 6:253–264

Moore WEC, Moore LVH (1994) The bacteria of periodontal diseases. Periodontol 2000 5:66–77

Moore WEC, Burmeister JA, Brooks CN, Ranney RR, Hinkelmann KH, Schieken RM, Moore LVH (1993) Investigation of the influences of puberty, genetics, and environment on the composition of subgingival periodontal floras. Infect Immun 61:2891–2898

Morhart R, Fitzgerald R (1980) Composition and ecology of the oral flora. In: Menaker L (ed) The biologic basis of dental caries. Harper & Row, Hagerstown, pp 263–277

Netuschil L, Reich E, Brecx M (1989) Direct measurement of the bactericidal effect of chlorhexidine on human dental plaque. J Clin Periodontol 16:484–488

Netuschil L, Reich E, Unteregger G, Sculean A, Brecx M (1998) A pilot study of confocal laser scanning microscopy for the assessment of undisturbed dental plaque vitality and topography. Arch Oral Biol 43:277–285

Netuschil L, Auschill TM, Sculean A, Arweiler NB (2014) Confusion over live/dead stainings for the detection of vital microorganisms in oral biofilms – which stain is suitable? BMC Oral Health 14:2

Newman HN, Wilson M (eds) (1999) Dental plaque revisited. BioLine, Antony Rowe Ltd, Chippenham

Obst U, Schwartz T, Volkmann H (2006) Antibiotic resistant pathogenic bacteria and their resistance genes in bacterial biofilms. Int J Artif Organs 29:387–394

Offenbacher S, Olsvik B, Tonder A (1985) The similarity of periodontal microorganisms between husband and wife cohabitants. J Periodontol 6:317–323

Okada M, Hayashi F, Nagasaka N (2001) PCR detection of 5 putative periodontal pathogens in dental plaque samples from children 2 to 12 years of age. J Clin Periodontol 28:576–582

Paster BJ, Olsen I, Aas JA, Dewhirst FE (2006) The breadth of bacterial diversity in the human periodontal pocket and other oral sites. Periodontol 2000 42:80–87

Percival RS, Challacombe SJ, Marsh PD (1991) Age-related microbiological changes in the salivary and plaque microflora of healthy adults. J Med Microbiol 35:5–11

Persson GR, Renvert S (2013) Cluster of bacteria associated with peri-implantitis. Clin Implant Dent Relat Res. doi:10.1111/cid.12052 [Epub ahead of print]

Petit MDA, van Steenbergen TJM, Timmerman MF, de Graaff J, van der Velden U (1994) Prevalence of periodontitis and suspected periodontal pathogens in families of adult periodontitis patients. J Clin Periodontol 21:76–85
Postgate JR (1969) Viable counts and viability. Meth Microbiol 1:611–628
Preus HR, Olsen I (1988) Possible transmittance of A. actinomycetemcomitans from a dog to a child with rapidly destructive periodontitis. J Periodont Res 23:68–71
Prosser JI (1999) Quorum sensing in biofilms. In: Newman HN, Wilson M (eds) Dental plaque revisited. BioLine, Antony Rowe Ltd, Chippenham, pp 79–88
Quirynen M, De Soete M, Dierickx K, van Steenberghe D (2001) The intra-oral translocation of periodontopathogens jeopardises the outcome of periodontal therapy. A review of the literature. J Clin Periodontol 28:499–507
Rams TE, Feik D, Slots J (1990) Staphylococci in human periodontal diseases. Oral Microbiol Immunol 5:29–32
Ritz HL (1967) Microbial population shifts in developing human dental plaque. Arch Oral Biol 12:1561–1568
Rönström A, Edwardsson S, Attström R (1977) Streptococcus sanguis and Streptococcus salivarius in early plaque formation on plastic films. J Periodontol Res 12:331–339
Saarela M, von Troil-Lindén B, Torkko H, Stucki A-M, Alahuusua S, Jousimies-Somer H, Asikainen S (1993) Transmission of oral bacterial species between spouses. Oral Microbiol Immunol 8:349–354
Salerno C, Pascale M, Contaldo M, Esposito V, Busciolano M, Milillo L, Guida A, Petruzzi M, Serpico R (2011) Candida-associated denture stomatitis. Med Oral Patol Oral Cir Bucal 16:139–143
Salvi GE, Fürst MM, Lang NP, Persson GR (2008) One-year bacterial colonization patterns of Staphylococcus arueus and other bacteria at implants and adjacent teeth. Clin Oral Implants Res 19:242–248
Saxton CA (1973) Scanning electron microscope study of the formation of dental plaque. Caries Res 7:102–119
Schlagenhauf U, Pommerencke K, Weiger R (1995) Influence of toothbrushing, eating and smoking on Dentocult SM Strip mutans test scores. Oral Microbiol Immunol 10:98–101
Slots J, Gibbons RJ (1978) Attachment of Bacteroides melaninogenicus subsp. assaccharolyticus to oral surfaces and its possible role in colonization of the mouth and of periodontal pockets. Infect Immun 19:254–264
Socransky SS, Haffajee AD, Cugini MA, Smith C, Kent RL Jr (1998) Microbial complexes in subgingival plaque. J Clin Periodontol 25:134–144
Sönju T (1987) Pellicle – formation, composition and possible role. Chapter 4. In: Thylstrup A, Fejerskov O (eds) Textbook of cariology. Munksgaard, p 46–55
Tanner A, Maiden MFJ, Macuch PJ, Murray LL, Kent RL Jr (1998) Microbiota of health, gingivitis, and initial periodontitis. J Clin Periodontol 25:85–98
ten Cate JM (2006) Biofilms, a new approach to the microbiology of dental plaque. Odontology 94:1–9
Theilade E, Theilade J (1970) Bacteriological and ultrastructural studies of developing dental plaque. In: McHugh (ed) Dental plaque. Livingstone, Edinburgh, pp 27–40
Theilade E, Wright WH, Jensen SB, Löe H (1966) Experimental gingivitis in man. II. A longitudinal clinical and bacteriological investigation. J Periodontol Res 1:1–13
Van der Velden U, Abbas F, Armand S, de Graaff J, Timmerman MF, van der Weijden GA, van Winkelhoff AJ, Winkel EG (1993) The effect of sibling relationship on the periodontal condition. J Clin Periodontol 20:683–690
Van Houte J, Green DB (1974) Relationship between the concentration of bacteria in saliva and the colonization of teeth in humans. Infect Immun 9:624–630
Van Palenstein Helderman WH (1981) Longitudinal microbial changes in developing human supragingival and subgingival dental plaque. Archs Oral Biol 26:7–12
Van Winkelhoff AJ, Boutaga K (2005) Transmission of periodontal bacteria and models of infection. J Clin Periodontol 32(Suppl 6):16–27
Voelkerding KV, Dames SA, Durtschi JD (2009) Next-generation sequencing: from basic research to diagnostics. Clin Chem 55:641–658
Von Troil-Lindén B, Alahuusua S, Wolf J, Jousimies-Somer H, Torppa J, Asikainen S (1997) Periodontitis patient and the spouse: periodontal bacteria before and after treatment. J Clin Periodontol 24:893–899
Wade WG (2013) Detection and culture of novel oral Bacteria. In: Jakubovics NS, Palmer RJ Jr (eds) Oral microbial ecology – current research and new perspectives. Caister Academic Press, Norfolk
Wimpenny JWT (1997) The validity of models. Adv Dent Res 11:150–159
Wimpenny J, Manz W, Szewzyk U (2000) Heterogeneity in biofilms. FEMS Microbiol Rev 24:661–671
Winterberg H (1898) Zur Methodik der Bakterienzählung [Concerning methods to count bacteria]. Z Hyg 29:75–93
Zaura E, Keijser BJ, Huse SM, Crielaard W (2009) Defining the healthy "core microbiome" of oral microbial communities. BMC Microbiol 9:259
Zaura-Arite E, van Marle J, ten Cate JM (2001) Confocal microscopy study of undisturbed and chlorhexidine-treated dental biofilm. J Dent Res 80:1436–1440
Ziegler NR, Halvorson HO (1935) Application of statistics to problems in bacteriology. IV. Experimental comparison of the dilution method, the plate count, and the direct count for the determination of bacterial populations. J Bacteriol 29:609–634

# The Microbiota of the Human Skin

**5**

Markus Egert and Rainer Simmering


## Abstract

The aim of this chapter is to sum up important progress in the field of human skin microbiota research that was achieved over the last years.

The human skin is one of the largest and most versatile organs of the human body. Owing to its function as a protective interface between the largely sterile interior of the human body and the highly microbially contaminated outer environment, it is densely colonized with a diverse and active microbiota. This skin microbiota is of high importance for human health and well-being. It is implicated in several severe skin diseases and plays a major role in wound infections. Many less severe, but negatively perceived cosmetic skin phenomena are linked with skin microbes, too. In addition, skin microorganisms, in particular on the human hands, are crucial for the field of hygiene research. Notably, apart from being only a potential source of disease and contamination, the skin microbiota also contributes to the protective functions of the human skin in many ways. Finally, the analysis of structure and function of the human skin microbiota is interesting from a basic, evolutionary perspective on human microbe interactions.

Key questions in the field of skin microbiota research deal with (a) a deeper understanding of the structure (species inventory) and function (physiology) of the healthy human skin microbiota in space and time, (b) the distinction of resident and transient skin microbiota members, (c) the distinction of beneficial skin microorganisms from microorganisms or communities with an adverse or sickening effect on their hosts, (d) factors shaping the skin microbiota and its functional role in health and disease, (e) strategies to manipulate the skin microbiota for therapeutic reasons.



M. Egert (✉)
Faculty of Medical and Life Sciences, Institute of Precision Medicine, Microbiology and Hygiene Group, Furtwangen University, Jakob-Kienzle-Str. 17, 78054 Villingen-Schwenningen, Germany
e-mail: Markus.Egert@hs-furtwangen.de

R. Simmering
Corporate Scientific Services, Henkel AG & Co. KGaA, Henkelstr. 67, 40589 Düsseldorf, Germany
e-mail: Rainer.Simmering@henkel.com



A. Schwiertz (ed.), *Microbiota of the Human Body*, Advances in Experimental Medicine and Biology 902, DOI 10.1007/978-3-319-31248-4_5

**Keywords**
Skin microbiota • *Bacteria* • *Archaea* • Fungi • *Actinobacteria* • *Firmicutes* • *Proteobacteria* • Homeostasis • Skin diseases • Skin defense and immunity • Cosmetics • Perspiration

## 5.1 Introduction

The human skin is one of the largest and most versatile organs of the human body. Owing to its function as a protective interface between the (mostly sterile) interior of the human body and the (unsterile) outer environment, it is densely colonized with a diverse and active microbiota. With respect to the number of microbial cells, the human skin is ranked fourth place among the various niches of the human body, that are colonized with microorganisms, outnumbered just by the human gastrointestinal tract, the oral cavity and the vagina (Wilson 2008). The skin microbiota is implicated in several severe skin diseases (e.g., acne, psoriasis, atopic dermatitis etc.) and plays a major role in skin wound infections. Several negatively perceived cosmetically relevant skin phenomena are linked with skin microbes, too, e.g. impure skin, dandruff and body odor. The skin microbiota is also of high relevance for the wide field of hygiene (clinical hygiene, production hygiene, personal hygiene etc.). Here, the microbiota of the human hands is of particular importance. However, apart from being a potential source of disease and contamination, the skin microbiota is also crucial for the protective function of the human skin, e.g. by contributing to the skin acid mantle, by triggering the skin immune system, or by preventing skin colonization with pathogenic microorganisms (colonization resistance). Finally, the analysis of structure and function of the human skin microbiota is also interesting from an evolutionary perspective on human microbe interactions.

In conclusion, the human skin microbiota, is without doubt of high importance for human health and well-being. The aim of this chapter is to sum up important progress in the field of human skin microbiota research achieved over the last years. The focus of the text is on the outer skin of humans (cutis, dermis); the microbiota residing on epithelia inside the human body (mouth, GI tract etc.) will be addressed in other chapters. In addition to this chapter, the reader is also referred to several excellent recent review articles on the structure and function of the human skin microbiota (Grice and Segre 2011; Kong 2011; Kong and Segre 2012; Rosenthal et al. 2011; Schommer and Gallo 2013). A brief historical view on the skin microbiota is given by Grice and Segre (2011); for more comprehensive overviews on the more classical literature (based mainly on culture-dependent studies) the reader is referred to the respective chapter in (Wilson 2008) or the reviews by Roth and James (1988) or Bojar and Holland (2002).

Key questions in the field of skin microbiota research deal with (a) a deeper understanding of the composition and distribution of the human skin microbiota in space (across the human body and the different niches of the human skin) and time (age of the host), (b) the distinction of resident and transient skin microbiota members, and (c) the distinction of beneficial (symbiotic) members of the skin microbiota or even beneficial microbial communities from (parasitic) microorganisms or communities with an adverse or sickening effect on their hosts. Finally, (d) abiotic and biotic factors shaping the skin microbiota composition in space and time and (e) strategies (preventive, therapeutic, cosmetic etc.) to manipulate the skin microbiota are of high medical but surely also of commercial interest.

## 5.2 The Human Skin as a Habitat for Microorganisms

The human skin is a complex organ with many important functions. As an **interface** between host and environment it provides a mechanical and biological barrier against chemical, physical and pathogenic threats. Furthermore, it participates in the process of thermoregulation and supports immunological functions. Melanogenesis is as well a process the skin is involved in, as it protects against the effects of UV-radiation. To be able to fulfill all these important functions, the skin comprises many different structures and a diversity of cells with different properties. Anatomically, the skin comprises two distinct compartments (Fig. 5.1): the **epidermis**, an avascular layer mainly composed of keratinocytes, and the **dermis**, a fibroblast-rich network of collagen and elastin fibers that provides the skin with strength and elasticity. The dermis also contains capillary and lymphatic vessels, which serve as the entry and exit portals for immune cells. Additional skin appendages such as hair follicles, sebaceous glands and sweat glands, as well as nerve endings are also found in the dermis. For a deeper understanding of the histology of the normal and healthy human skin, the reader is referred to (Urmacher 1990).

Due to their very versatile physiology, microorganisms are able to colonize many of the different niches (micro-ecosystems) that are realized on human skin. Some areas of the skin are a rather **sebaceous** environment, resulting in a microbiota dominated by lipophilic bacteria such as propionibacteria. These areas are found on the scalp, the forehead, the neck and partly on the upper part of the back. Other parts of the body are **moist** and **warm**, like the armpit, the genital area, and the feet. Finally, there areas which are relatively **dry**, such as the forearms or the legs and lower part of the back. These differences are caused by the uneven distribution of sweat and sebaceous glands over the body.

As mentioned above, an important role of the human skin is the regulation of **homeostasis**, i.e. body temperature and water content (Percival et al. 2012). Organ failure due to denaturation of proteins and subsequent cell death is caused by perpetual elevation of the body core temperature above 40 °C. Therefore, regulation of body temperature represents a fundamentally important process (Wilke et al. 2007). This regulation is achieved by **perspiration**, i.e. the secretion of sweat. Sweat is secreted by sweat gland, i.e. specialized exocrine glands that are appendages of the skin (Fig. 5.1). Sweat glands can be categorized into eccrine and apocrine glands. Sweat

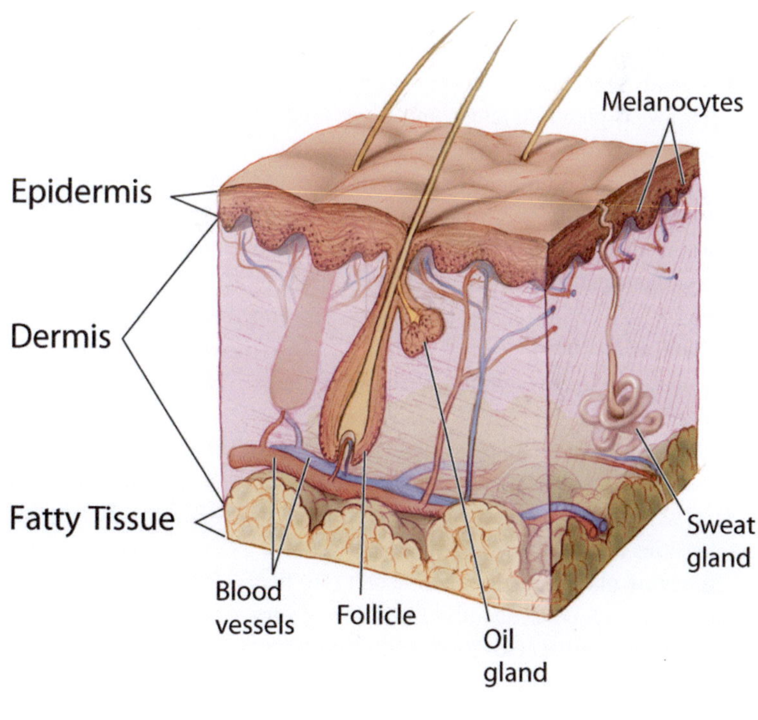


**Fig. 5.1** Layers of the human skin and associated glands and vessels. Microbial life was long-believed to be restricted to the epidermis and its appendages (hair follicles and glands). However, recent studies suggest skin microorganisms also to be present in deeper layers, where they can interact with the skin's immune system (Picture credit: Don Bliss; National Cancer Institute, USA)

glands exhibiting attributes of both types are named apoeccrine sweat glands. The distribution of these cutaneous glands varies over the body. In addition, each gland type is regulated by different stimuli and has different functions (Wilke et al. 2007; Noël et al. 2012).

**Eccrine sweat glands** are the most abundant sweat glands. On average, 100–200 glands per $cm^2$ are distributed virtually all over the body surface. The palms and soles exhibit higher densities of about 600 glands per $cm^2$. In contrast, body parts such as the lips and nail bed are depleted of eccrine glands (Noël et al. 2012). Eccrine sweat glands are effective from birth on in executing their main function, i.e. thermoregulation. Thermoregulatory perspiration is affected by environmental parameters, such as temperature, humidity, skin and body temperature in general, but also by physical fitness, circadian rhythm and the menstrual cycle. While temperature is the major onset of eccrine glands, these glands are also activated by pain, stress, fear, and anxiety resulting in emotional sweating. Digestion is also speculated to induce eccrine sweating. However, the underlying mechanisms are not well understood to date (Wilke et al. 2007).

Another crucial function of eccrine sweat is the prevention of bacterial colonization and growth through acidification of the skin surface (Grice and Segre 2011). Sweat secreted from eccrine sweat glands is clear and composed of mainly water containing sodium and potassium salts as well as amino acids, sugars, lactate and glycoproteins (Kelly and Wood 2010). Its exact composition differs depending on hormonal activity, physical condition, acclimatization to environmental conditions as well as secretion rate (Noël et al. 2012).

**Apoeccrine sweat glands** were first introduced by (Sato et al. 1989). They are far less abundant on skin than eccrine glands. About 2,000 of them are distributed around the eyes and ears as well as in the breast skin, while the highest densities are reported for axillae and groin.

**Apocrine glands** release their secret into hair canals, and are consequently limited to hairy body surfaces (Wilke et al. 2007; Kelly and Wood 2010; Noël et al. 2012). They are not activated until the onset of puberty and androgen stimulation (Wilke et al. 2007; Noël et al. 2012). Following activation during puberty, apocrine glands are then stimulated by hormones. Apocrine sweat appears milky and viscous, and is known to comprise lipids, lactate, nitrogen, electrolytes, steroids, proteins and vitamins besides other ions (Noël et al. 2012; Fredrich et al. 2013). It is also suspected to contain pheromones (Wilke et al. 2007). However, the exact composition of apocrine secretion has yet to be elucidated due to the lack of pure samples. In addition, while apocrine glands are suspected to be involved in emotional sweating, their distinct function remains unidentified (Wilke et al. 2007; Fredrich et al. 2013). Whereas eccrine sweat glands contribute only slightly to human body odor, apocrine sweat is at least well accepted to contain substances that can be transformed into odorous molecules upon bacteriolysis (Wilke et al. 2007; Kelly and Wood 2010; Grice and Segre 2011; Noël et al. 2012; Fredrich et al. 2013).

An important feature of skin and mucosa is – in contrast to abiotic surfaces – the ability to continuously (re)generate new cells and to create response reactions which provide protection against microbial infection and degradation. This innate immune system has to be considered as an additional and important factor influencing the microbial equilibrium on human skin. Langerhans cells in the epidermis as well as dendritic cells, macrophages, mast cells, T and B cells, plasma cells and natural killer cells in the dermis participate in immune responses within the skin. Many cell types that permanently reside in the skin produce anti-microbial peptides, including keratinocytes, sebocytes, eccrine glands, and mast cells (Schauber and Gallo 2008).

Thus, an important part of the **antimicrobial defense system** of the human skin constitutes of small cationic peptides, i.e. human ß-defensins hBD-1, hBD-2, hBD-3, hBD-4 (García et al. 2001; Harder et al. 2001; Schauber and Gallo 2008), of the human cathelicidin LL-37 (Frohm et al. 1997), of antimicrobial enzymes like lysozyme and RNAse 7 (Harder and Schroder 2002) and of several other molecules exhibiting antimicrobial potential (for a review see Braff et al. 2005).

The active form of Cathelicidin LL-37 is derived from the precursor protein hCAP18 by cleavage by serine proteases. It consists of 37 amino acids, exhibits an α-helical form and is active against bacteria, fungi and viruses (Braff and Gallo 2006).

**ß-defensins** are small peptides (4–5 kDa) with a characteristic set of three disulfide bonds. In general, hBD-1 is regarded as being constitutively expressed in the epithelium, while hBD-2 and hBD-3 are only induced by inflammation. In acne lesions, a strong induction of hBD-2 was observed in highly inflammated pustules, while hBD-1 was only moderately expressed with the strongest signal in the papules (Philpott 2003). In contrast, hBD-2 was not upregulated in atopic dermatitis (Ong et al. 2002), while both hBD-2 and hBD-3 were expressed in inflamed psoriatic lesions (Harder and Schroder 2002; Nomura et al. 2003). In keratinocytes exposed to *Staphylococcus aureus*, the expression of hBD-2 was strongly induced, while hBD3 and LL-37 showed only moderate, and hBD-1 virtually no induction (Midorikawa et al. 2003). *Malassezia furfur* was shown to induce the expression of hBD-2 via protein kinase C, but not of hBD-1 in keratinocytes (Donnarumma et al. 2004). RNAse 7 was found to be induced in keratinocytes by contact with heat-inactivated cells of bacterial pathogens like *Pseudomonas aeruginosa*, *Staphylococcus aureus*, *Escherichia coli* and *Streptococcus pyogenes* (Harder et al. 2001).

These antimicrobial peptides represent one element, among others, of the **innate immune system** of the skin. Their role is by far not only restricted to the direct inactivation of skin microorganisms but also includes stimulation of further cellular reactions. As an example, a complex cascade involving IL-6, IL-10 and other cytokines is activated by cathelicidin LL-37 via cell-surface receptors (for an overview see Schauber and Gallo 2008).

Among the constitutive properties of skin that help preventing colonization and infection by microorganisms are finally its relatively low temperature and the acidic pH (Grice and Segre 2011).

## 5.3 Structure and Variability of the Healthy Skin Microbiota

The skin microbiota of humans comprises bacteria, fungi (mostly yeasts), viruses and – which is only known since recently – also archaea, i.e. all three **domains** of life. In addition, higher, parasitic eukaryotes (mostly arthropods) can occur (e.g., mites) but will not be discussed here, as they do not represent microorganisms sensu stricto. Recent investigations on the **structure**, i.e. the species inventory, of the human skin microbiota were significantly influenced by the application of high throughput next generation sequencing technologies, which allow the analysis of millions of nucleic acid sequences (e.g., 16S rRNA or 18S rRNA genes) in a single study. It is an advantage that molecular methods also detect microbes that are difficult to culture or even have not been cultured yet. However, they are usually inappropriate to differentiate living from dead microbes and can be affected by well know biases, such as incomplete nucleic acid extraction from the samples or discrimination of sequences types due to mismatches with the primers used for PCR amplification (Forney et al. 2004). Nonetheless and interestingly, (older) cultivation-based and (more recent) large scale molecular studies on the skin micobiota yielded rather similar results with regard to the dominant groups of microbial species on human skin.

### 5.3.1 The "normal" Skin Microbiota of Healthy Adults

In order to define a "normal" skin microbiota healthy adults were studied. An adult and healthy human being is approximately colonized by $10^8$–$10^{10}$ skin microbes in total, which are distributed unevenly across the different niches of the human skin. **Cell numbers** range from approximately $10^2$ cm$^{-2}$ (fingertips, back) to $10^6$ cm$^{-2}$ (forehead, axilla). Owing to its physico-chemical conditions (see previous paragraph), the skin is typically colonized by **mesophilic**, **xerophilic**, **acidophilic**, **osmotolerant**, and **facultative aerobic** microorganisms. However, depending on the respective niche, also microbes with other physiological traits can occur. While it was long believed that on healthy skin microbial life is restricted to the **epidermis** and appendages such sebaceous and sweat glands (see Fig. 1 in (Grice and Segre 2011)), recent analyses (Nakatsuji et al. 2013) suggest microbial life also to occur in deeper skin layers, i.e. the **dermis** and underlying **fatt tissue** (Fig. 5.1). This finding is of high importance from an immunological point of view, because it suggests direct communication between host and microbial cells in a tissue previously thought to be sterile.

#### 5.3.1.1 Bacteria

*Bacteria* represent by far the most abundant and best studied group of living microorganisms on healthy human skin. The vast majority of them are affiliated with **three phyla** (typical and abundant genera in brackets): *Actinobacteria* (*Corynebacterium*, *Propionibacterium*, *Micrococcus*, *Brevibacterium*), *Firmicutes* (*Staphylococcus*, *Streptococcus*) and *Proteobacteria* (*Acinetobacter*, *Methylobacterium*). However, recent studies using high throughput sequencing technologies have shown, that the skin microbiota is highly diverse and comprises members affiliated with more than 25 different phyla, albeit mostly at low abundances. Such investigations also revealed that the composition of the skin microbiota is highly individual and changes significantly over time.

In a pioneering – and according to today's standards now small scale – study, Gao and colleagues (2007) sampled the superficial volar left and right **forearms** in six healthy adult subjects by means of swabbing. Based on approximately 1,200 cloned and sequenced partial 16S rRNA genes they identified 182 operational taxonomic units ("species") belonging to 91 genera and 8 phyla. On average, 48 species were detected per human individual. *Actinobacteria*, *Firmicutes*, and *Proteobacteria* accounted for ~95 % of the clones. Interestingly, and in sharp contrast to many other microbial ecosystems on earth, 85 % of the sequences corresponded to known and yet cultivated species. Analysis of 817 clones obtained 8–10 months later from four subjects showed two additional phyla, 28 additional genera and 65 additional species. Only four (3.4 %) of the 119 genera (*Propionibacterium*, *Corynebacterium*, *Staphylococcus*, *Streptococcus*) were observed in each subject tested twice, albeit these genera represented 54.4 % of all clones.

In a subsequent **landmark study**, Grice and colleagues (2009) analyzed more than 112,000 nearly full length bacterial 16S rRNA gene sequences obtained from **20 different body sites** of 10 healthy adults, half of them were sampled a second time 4–6 months after the first sampling. Nineteen bacterial phyla were detected, but most sequences were assigned to just four phyla: *Actinobacteria* (52 %), *Firmicutes*, (24 %), *Proteobacteria* (17 %), and *Bacteroidetes* (6 %). Of the 205 identified genera represented by at least five sequences, three were associated with more than 62 % of the sequences: Corynebacteria (23 %, *Actinobacteria*), propionibacteria (23 %; *Actinobacteria*), and staphylococci (17 %; *Firmicutes*). Interestingly, the diversity and temporal stability of the microbial community were dependent on the specific characteristics of the skin site. Sebaceous sites were dominated by propionibacteria and staphylococci, Corynebacteria and (to a lower extent) staphylococci predominated in moist sites. A mixed population of bacteria resided in dry sites, with a greater prevalence of *ß-Proteobacteria* and *Flavobacteriales*. In view of the fact that Gram-negative bacteria are usually restricted to rather moist habitats, the last finding was unexpected. Species diversity, measured as Shannon-diversity index, was side-dependent; it was lowest for the back, retroauricular crease and toe web space samples and highest for the popliteal fossa, plantar heel and antecubital fossa samples. Longitudinal (i.e., temporal) stability of the skin microbiota was also site-dependent: it was high for protected sites such as the nares and the external auditory canal and low for samples from more exposed sites, such as buttock and popliteal fossa. In summary, the study impressively proved the strong influence of the respective skin niche on the composition of the skin microbiota (Fig. 5.2), which is, in fact, more dependent on the investigated skin site than on the individual (Grice and Segre 2011). Rightly, the authors considered their study as an important baseline for further studies on the role of the skin microbiota in human health and disease.

Another landmark study was conducted by Fierer and colleagues (2008) on the microbiota of **human hands**. Here, more than 350,000 partial 16S rRNA gene sequences were obtained from left and right hands (palmar surface) of 27 healthy men and 24 healthy women. Across all samples they determined more than 4,700 bacterial phylotypes ("species"); on average, every hand was colonized by 158 different phlyotypes. More than 25 phyla were detected, however, *Actinobacteria*, *Firmicutes*, and *Proteobacteria* accounted for 94 % of the sequences, and again, the most abundant genera were *Propionibacterium* (31.6 % of all sequences), *Streptococcus* (17.2 %), *Staphylococcus* (8.3 %); *Corynebacterium* (4.3 %;) and *Lactobacillus* (3.1 %). These genera were found on virtually all palm surfaces sampled. This pronounced microbial diversity – being in the order of magnitude of the human intestinal tract – was not expected. However, it can easily be explained with the function of the human hand as central "grabbing organ", that daily gets in contact with many unsterile surfaces and is exposed to many factors with a strong influence on microbial community composition.

The authors also observed pronounced intra- and interpersonal variation in bacterial community composition. Hands from the same individual shared only 17 % of their phylotypes, with different individuals sharing only 13 %. Interestingly, women had a significantly higher diversity than men, and community composition was significantly affected by handedness, time since last hand washing, and an individual's sex. The resident and transient microbiota of the human hand is of high practical relevance from an hygienic point of view. A very recent review on this topic, albeit with a focus on molecular studies, is given by (Edmonds-Wilson et al. 2015).

A fascinating study on the microbiota in human **belly buttons** (Hulcr et al. 2012) showed that also this "frequently overseen" habitat is microbiologically as diverse as other skin habitats. On average, 67 bacterial phylotypes were found per belly button. However, the communities were strongly dominated by a few taxa: only six phylotypes occurred on 80 % of all humans. Abundant phylotypes were affiliated with staphylococci, corynebacteria, and several genera of *Actinobacteria* (e.g., *Micrococcus*) and *Clostridiales* (e.g., *Anaerococcus*, *Finegoldia*, *Peptinophilus*), bacilli, and, to a lesser extent, *Gamma-Proteobacteria* (e.g., *Acinetobacter*). The authors hypothesize that the macroecological concept of "oligarchs" can be applied to skin microbiomes. "Oligarchs" are groups of organisms that are closely related and dominate the community composition in many, but (in contrast to "core species") not all samples. They represent the organisms that are evolutionary best adapted to the respective habitat.

#### 5.3.1.2 Fungi, Archaea and Viruses

In comparison to bacteria, relatively little is known about other members of the human skin microbiota, such as fungi, archaea and viruses. Cultivation-based and earlier molecular studies agreed that the human skin **fungal microbiota** (the skin mycobiome) is dominated by yeasts, in particular of species affiliated with the genus *Malassezia*. Findley and colleagures (2013) analyzed fungal communities of 14 skin sites in 10 healthy adults by means of Next Generation Sequencing (NGS) of PCR-amplified 18S rRNA genes and ITS regions. Eleven core-body and arm sites were dominated by fungi of the genus *Malassezia*, with only species-level classifications revealing fungal-community composition differences between sites. By contrast, three foot sites showed a considerably high fungal diversity, comprising also other genera (in addition to *Malassezia*) such as *Aspergillus*, *Cryptococcus*, *Rhodotorula*, *Epicoccum* and others, and a lower stability over time. Fungal

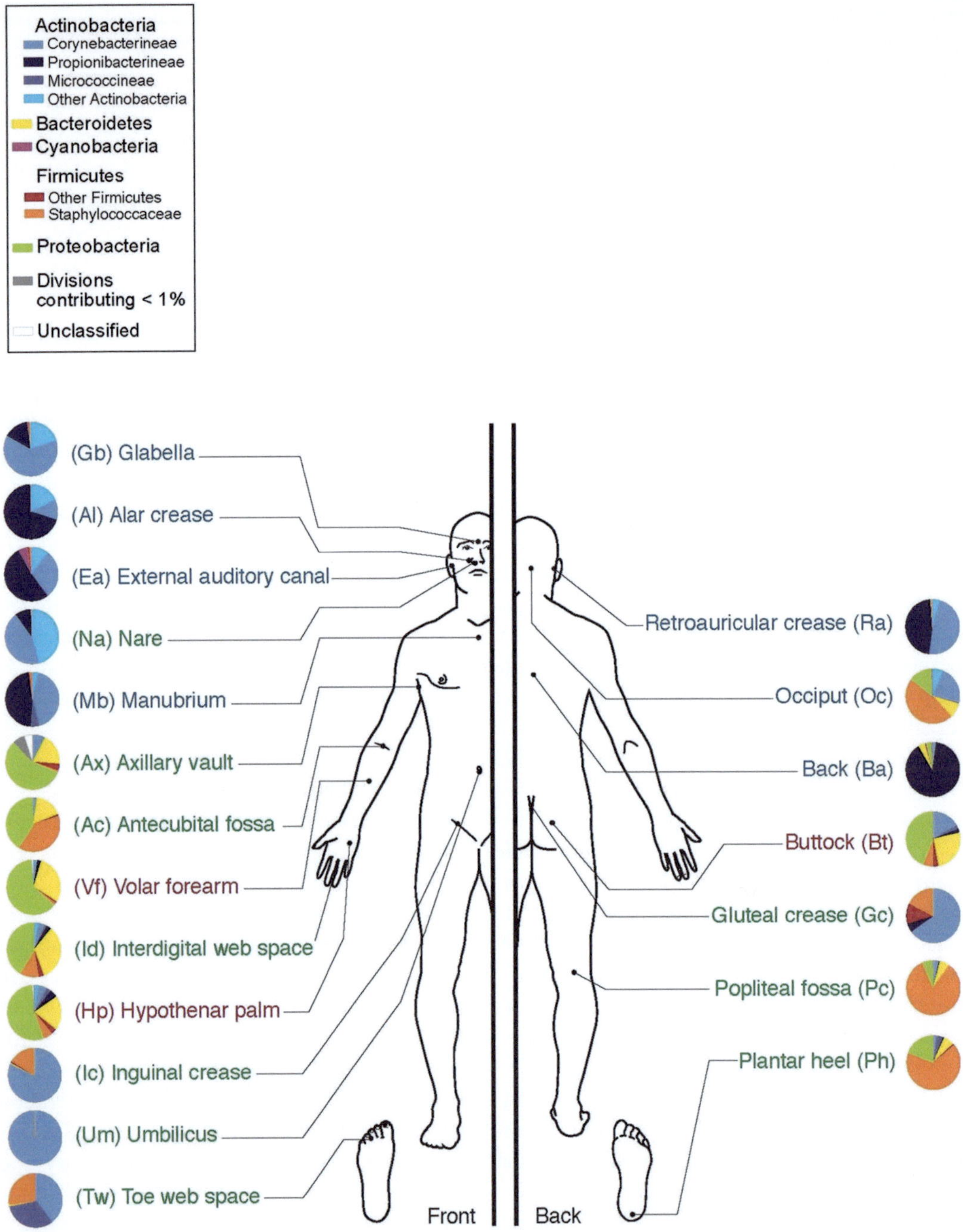


**Fig. 5.2** Diversity map of major bacterial groups on phylum and family-level from a human individual. The figure is based on data from the landmark study by (Kong and Segre 2012). Some results are unusal, such as the dominance of proteobacteria under the axilla, but can be explained with the fact that data from a single individual are shown (Picture credit: Darryl Leja, National Human Genome Research Institute, USA; http://www.genome.gov/dmd/img.cfm?node=Photos/Graphics&id=85320)

communities were also strongly influenced by skin topography, but – in comparison to bacterial skin communities –, less clearly by skin physiology. Interestingly, *Malazessia* species were by far not the most abundant fungi in a study on the fungal microbiota on healthy scalps and scalps with dandruff (Park et al. 2012). Based on PCR-NGS-based analyses of 26 rRNA genes, *Acremonium* spp. represented ca. 62 % of all sequences on healthy scalps while *Malazessia* spp. did only 0.07 %.

Maybe with exception of the human intestinal tract, the role of ***Archaea*** in the human microbiota is still very unclear and under debate (Horz and Conrads 2010; Horz 2015). Until recently, they could not be detected in human skin samples, neither by cultivation nor by PCR-based methods (Gao et al. 2008), and consequently, their presence was questionable (Grice and Segre 2011). However, a recent report (Probst et al. 2013) suggested that archaea can represent more than 4 % of the human prokaryotic skin microbiota. Phylotypes detected by PCR and FISH were mostly affiliated with *Thaumarchaeota* and to a lesser extent also with *Euryarchaeota*. The physiological role of these skin archaea is absolutely unclear, a role in skin ammonia metabolism is suggested, following the function of *Thaumarchaeota* in many environmental ecosystem.

While it is widely accepted that the human skin, in particular the skin of the human hands, can transmit **viruses** (Julian et al. 2010), very little is known about a potentially commensal or resident viral microbiota of the human skin, i.e. the skin virome. In a landmark study, Foulongne and colleagues (2012) used a highthroughput metagenomic approach to study the skin microbiota of five healthy patients and one patient with skin cancer. Significant proportions (up to more than 87 %) of the obtained DNA sequences per patient represented DNA viruses, mostly affiliated with polyomaviruses, papillomaviruses and circoviruses, even on healthy appearing skin. The physiological role of these viruses, partly representing new species, in human health and disease still remains to be elucidated.

### 5.3.2 Factors Shaping the Composition of the Skin Microbiota

Several factors have been identified that influence the composition of the skin microbiota of the human body in space and time. In general, these factors can be divided into **intrinsic** (**host**) and **extrinsic** (**environmental**) factors (Grice and Segre 2011). Table 5.1 provides a selection of recent studies dealing with various factors that shape the overall composition of the skin microbiota. A more detailed view on the relationship between certain skin diseases and changes in skin microbial community composition is given in paragraph 5.4.2. An up-to-date review on host factors that interact with the human skin microbiota can be found in (SanMiguel and Grice 2015).

In addition to the factors mentioned in Table 5.1, several others are likely to influence the composition of the human skin microbiota, such as diet, climate, solar/UV-radiation, occupation, non-topical use of antibiotics, stress etc., however, appropriate studies are still missing.

Finally, **interactions** among the different members of the skin microbiota are another exciting, yet also still poorly investigated factor with influences on the overall composition of the skin microbiota. Staphylococci can be regarded as the best investigated bacteria in this respect. *Staphylococcus epidermidis*, a typical commensal of the human skin, was shown to produce antimicrobial peptides, e.g. phenol-soluble modulins, that selectively inhibit the growth of skin pathogens such as *Staphylococcus aureus* and group A streptococci (Cogen et al. 2010). In addition, Iwase and colleagues (2010) showed that some commensal strains of. *S. epidermidis* can protect their host from colonization with *S. aureus* by means of a serine protease (*Esp*) that destroys biofilms, for instance in the anterior nares. Much less in known about interactions including fungi and/or viruses, although – facing the situation in the human gastrotintestinal tract (Minot et al. 2011) – it appears likely that for

**Table 5.1** Summary of various host (h) and environmental (e) factors shaping/influencing the composition of the human skin microbiota

| Factor | Effects/observations | References |
|---|---|---|
| Skin site[h] | Differences in physiological parameters at different skin sites significantly shape skin microbiota, leading to a higher intrapersonal than interpersonal variability | Findley et al. (2013), Grice et al. (2009) |
| Sex[h] | Skin microbiota of males and females differs significantly, presumably due to physiological and/or behavioral reasons | Fierer et al. (2008), Giacomoni et al. (2009) |
| Early age[e] | Diversity of skin microbiota increases within first year of living; influence of delivery mode disappears | Capone et al. (2011) |
| Delivery mode[e] | Skin flora of newborns delivered vaginally is dominated by vaginal bacteria; cesarean section leads to skin microbiota dominated by typical skin bacteria. Implications for early infections of newborns suspected | Dominguez-Bello et al. (2010) |
| Personal hygiene/use of cosmetics[e] | Axillary cosmetics modify the microbial community and can stimulate odor-producing bacteria | Callewaert et al. (2014) |
| Life style/sports[e] | Skin to skin contact shaped skin microbiota composition of roller derby players | Meadow et al. (2013) |
| Hand washing[e] | Hand washing altered relative abundances of bacterial groups on human hands but not overall bacterial diversity | Fierer et al. (2008) |
| Genetic predisposition[h] | Mutations in the Filaggrin gene disturb the skin barrier function and are correlated with atopic dermatitis and significant changes in skin microbiota composition | McAleer and Irvine (2013) |
| Immune status[h] | Patients with primary immunodeficiencies show a significantly altered skin microbiota | Oh et al. (2013) |
| Skin diseases[h] | Most skin diseases go along with significant changes in skin microbiota composition. It is largely unclear whether this is cause or effect of the respective disease. Antibiotic treatments can sometimes alleviate disease symptoms | See paragraph 5.4.2 and studies cited therein |
| Underlying (non-skin) diseases[h] | Patients with diabetes were more likely to carry *S. aureus* even on their forearms; an altered skin microbiota might play a role in wound-infections | Redel et al. (2013) |
| Geography[e] | The microbiota of long uncontacted people might serve as a basis for a "natural" microbiota, not/less affected by industrialization | Clemente et al. (2015) |

instance bacteriophages also shape the composition of the skin microbiota. Interactions between fungi and bacteria are suggested by Findley and coworkers (2013). For human feet, they found negative correlations between *Actinobacteria* and feet *Ascomycota* and *Basidiomycota* and positive correlations with *Firmicutes* and *Proteobacteria*.

The previous brief discussion of microbe-microbe interactions leads to a more functional view of the human skin microbiota in human health and disease (Paragraph 5.4). A deeper knowledge of **microbe-microbe interactions** is also the prerequisite for rational manipulation (prevention, curation) strategies of skin microbiota related diseases, in particular probiotic strategies, i.e. strategies involving living (beneficial) bacteria (Paragraph 5.5). A promising pilot study (Harris et al. 2009) was performed with amphibians: Addition of violacein-producing *Janthinobacterium lividum* to the skin of frogs prevented them from the (fatal) fungal skin disease chytridiomycosis.

## 5.4 Functional Aspects of the Human Skin Microbiota

Due to the rapid progress in DNA sequencing technologies, recent years have brought considerable progress in unraveling the diversity of the human skin microbiota in many niches of the human body as well as major influencing factors. In comparison, considerably less is known about the **functionality** of the human skin microbiota and its role in human **health** and **disease**, and for human well-being. For recent review articles on that particular topic the reader is referred to (Rosenthal et al. 2011; Sanford and Gallo 2013; Zeeuwen et al. 2013).

### 5.4.1 Protective Functions of the Human Skin Microbiota

Nowadays it is well accepted that colonization of the human skin with a normal, balanced or healthy microbiota is in general beneficial for humans, because it protects them from skin infections and other skin diseases/disorders, a phenomenon also known as "colonization resistance". However, in view of pronounced temporal, intrapersonal as well as interindividual variations in microbial diversity, the definition of a "healthy", "normal" or "balanced" skin microbiota (i.e. as status of microbiological skin **homeostatis**) is difficult. Moreover, assigning distinct beneficial functions to individual members of the complex skin microbiota, comprising also yet-uncultured species, is even more challenging. Nevertheless, recent years have also brought considerable progress here.

The skin "**acid mantle**" of ~pH5 represents an important line of defense against pathogenic microorganisms. Facultative anaerobes, such as *Propionibacterium acnes*, reside under the anaerobic conditions of sebaceous glands, where they release free fatty acids from sebum onto the skin, which subsequently contribute to the acidic pH (Grice and Segre 2011). The capability of *Staphylococcus epidermidis* to produce antimicrobial peptides against skin pathogens such as *S. aureus* and group A streptococci was already discussed above.

Close interactions between the microbiota and the human **immune system** are well known from the human intestinal tract (recently reviewed by (Min and Rhee 2015). There, the presence of a gut microbiota is regarded important for the development and regulation of innate and adaptive immune systems and maintenance of gut homeostasis. The same obviously applies to the microbial human skin (Belkaid and Segre 2014; Nakamizo et al. 2015).

For instance, Lai and co-workers (2009) showed that the skin microbiota can modulate Toll-like-receptor (TLR-) dependent cutaneous inflammatory responses. *S. epidermidis* cell wall-derived lipoteichoic acid was shown to prevent skin injury-based skin inflammation by inhibition of inflammatory cytokines release from keratinocytes as well as TLR2-based immune-responses caused by RNA, deliberated from the injured skin cells. So far, skin bacteria were rather believed to cause than to reduce skin inflammation. In addition, it was shown that *S. epidermids* can augment skin defense mechanisms against infection by enhancing gene expression of antimicrobial peptides such as human beta defensins (Lai et al. 2010). Finally, it was shown that *S. epidermids* can also tune the function of resident skin T-lymphocytes and thereby contribute to protective immunity against skin pathogens (Naik et al.

2012). For instance, germ-free mice produced less interleukin-17A and interferon-γ than specific-pathogen-free mice. However, monoassociation of the skin of germ-free mice with *S. epidermidis* restored the production of IL-17A in skin T-cells. While the skin of germ-free mice reacted immunologically abnormally against an infection with *Leishmania major*, a protozoan parasite, monoassociation with *S. epidermidis* rescued protective immunity. Notably, these effects were not dependent on the intestinal flora of the investigated mice (Naik et al. 2012).

In summary, these results suggest that skin commensals such as *S. epidermidis* and *P. acnes* are important drivers and amplifiers of **human skin immunity** (Nakamizo et al. 2015) and a fine example that the human body comprises several niches (in addition to the intestinal tract), were immunosurveillance systems are locally fine-tuned by a commensal microbiota (Belkaid and Naik 2013). A nice overview on skin commensals as "**host guardians**" with a special emphasis on the two probably best studied skin commensals so far (*P. acnes* and *S. epidermidis*) was recently given by (Christensen and Brüggemann 2014). Consequently, disturbance of microbiological skin homeostatis might lead to skin dysfunctions and diseases.

### 5.4.2 Role in Skin Disorders and Skin Diseases

Many human skin disorders and diseases have been linked with changes in skin microbial community composition, however, in most cases it is still not clear, whether these observed changes are **cause or effect** of the underlying disease, which is, however, an important basis to chose an adequate therapy.

In the case of **acne** (Acne vulgaris), a chronic inflamatory disease of the pilosebaceous unit which can be regarded as the most prevalent human skin disease on earth, *Propionibacterium acnes* is for long known as the primary disease-associated bacterium. Despite the pronounced diversity of the human skin microbiota, it has been shown that healthy and diseased hair follicles are virtually exclusively colonized by *P. acnes* (Bek-Thomsen et al. 2008), which underlines the importance of this single species in acne pathology. Consequently, complete genome analysis of *P. acnes* contributed much to the understanding of the variety of virulence factors causing acne, e.g. enzymes such as hyaluronidases, lipases and proteases (Brüggemann et al. 2004). However, it also became clear that the virulence of different *P. acnes* strains can differ considerably, despite high identities on the genome level, suggesting more pronounced differences in gene expression among these strains (Brüggemann 2005). Notably, the most important basic pathomechanism of acne is hormone-induced increased sebum production, providing *P. acnes* with ideal living conditions in an anaerobic and lipid-rich environment.

In the case of **Psoriasis** and **Atopic dermatitis**, genetic and environmental patho-factors appear to be involved (Schommer and Gallo 2013). Gao and co-workers (2008) showed higher frequencies of *Firmicutes* and lower frequencies of *Actinobacteria*, in particular of propionibacteria, in psoriatic lesions compared to normal skin stretches of patients and of skin from healthy patients. Based on the analysis of 51 triplet samples from diseased, unaffected and healthy (control) skin, Alekseyenko and coworkers (2013) concluded that psoriasis induces physiological changes both at the lesion site and (!) at the systemic level. Lesions were characterized by a reduced microbial species diversity. Psoriasis patients were characterized by higher combined relative abundances of the major skin genera *Corynebacterium*, *Propionibacterium*, *Staphylococcus*, and *Streptococcus* (*Firmicutes-Actinobacteria* microbiota cutaneotype), while genera such as *Cupriavidus*, *Methylobacterium*, and *Schlegelella* were significantly less abundant (*Proteobacteria*-microbiota cutaneotype). Although the underlying, selecting patho-factors remain unclear, the authors conclude that these findings may have important diagnostic, preventive, and potentially therapeutic implications.

**Atopic dermatitis** is characterized by an impaired barrier function of the skin, leading to increased bacterial colonization and more frequent infections with *S. aureus*. Lower production of skin antimicrobial peptides and a disturbed skin cornification due to a mutation-based disturbed filaggrin production have been identified as potential pathological reasons here (Schommer and Gallo 2013). AD-patients also showed a significantly altered skin microbiota at sites of disease predilection, in particular higher shares of staphylococci (Kong et al. 2012; Seite et al. 2014). These differences could for instance be reversed by an emollient treatment (Seite et al. 2014). Interestingly, *Stenotrophomonas* species were significantly more abundant in the communities of patients that responded to emollient therapy, suggesting a possible role of this genus in restoration of the skin microbiota in patients with AD.

**Rosacea** and **Serborrheic dermatitis** are skin disease, where non-bacterial members of the skin microbiota are thought to play a major role, i.e. *Demodex* mites and *Malassezia* fungi, respectively, and take favor of an (immunological and/or prokaryotic) dysbiosis of the skin ecosystem (Schommer and Gallo 2013). In the case of Rosacea, Microbiota-associated changes on the skin (and simultaneously the small intestine!) have recently been summarized elsewhere (Picardo and Ottaviani 2014). Serborrheic dermatitis of the scalp is also known as dandruff, affects ~50 % of the global population and is mainly caused by *M. restricta* and *M. globosa*. So far unknown factors might switch them into a pathogenic state, in which they produce/secret irritant fatty acids leading to skin hyperproliferation and scaling (Schommer and Gallo 2013).

The skin microbiota also plays an important role in **wound healing**. A culture-independent pilot study (Hannigan et al. 2014) of open fracture wounds showed pronounced differences in microbial community composition between wound center (significantly enriched with pseudomonds) and adjacent skin, which was characterized by a normal skin microbiota. These differences disappeared during healing. In general, the microbiota of wounds appears to be depended on several factors such as wound type (blunt, penetrating, chronic, acute), affected part of the body, and underlying (chronic) diseases, such as diabetes (Tomic-Canic et al. 2014). Interestingly, using germ-free mice, Canesso and co-workers (2014) could show that in the absence of a commensal microbiota, skin wound healing is accelerated and scarless, partially because of reduced accumulation of neutrophils, increased accumulation of alternatively activated healing macrophages, and better angiogenesis at wound sites.

In addition to the above-mentioned (severe) skin disorders and diseases, the skin microbiota also affects several **cosmetic skin problems**, such as impure skin, sensitive skin, dandruff, or body odor. In fact, many of these problems are just milder/less severe forms of the above-mentioned diseases. Over the last years, considerable progress has been made in the field of **body odor formation**, recently reviewed in (James et al. 2013). In the armpit, which represents one of the most densely colonized parts of the human skin, the resident microbiota metabolizes odorless sweat to malodorous compounds using hydrolytic enzyme activities, such as aminoacylase and c-s lyase. While it was long believed that mainly corynebacteria produce body odor, recent cultivation-independent studies (Troccaz et al. 2015) suggest that also other groups are involved here, which is corroborated by studies on anaerococii (Fujii et al. 2014) and staphylococci (Egert et al. 2013, 2014; Bawdon et al. 2015). As also shown for other skin habitats, the composition of the armpit microbiota is gender-specific (Egert et al. 2014; Troccaz et al. 2015). Moreover, it is characterized by an asymmetric (left vs. right) activity, which was shown using a differential 16S rRNA gene vs. 16S rRNA sequencing approach, that allows discrimination of active from less active microbial populations (Egert et al. 2011). Microbially-caused body odor formation and suitable prevention and treatment strategies are not only an economically, but also medically important field of research, as strong body odor (bromidrosis) is a pathological phenomenon (Mao et al. 2008). Moreover, body odor triggers a person's attractiveness for mosquitoes and might play a role in the transmission of infectious diseases such as malaria (Verhulst et al. 2011).

## 5.5 Manipulation of the Human Skin Microbiota

The human skin microbiota is a dynamic environment, which changes over time and space. Latest research indicates that the differences between subjects and/or between microbial colonized regions are higher than the inter-individual differences over time (Grice and Segre 2011). However, besides the "natural" dynamics occurring in those ecosystems, also artificial, i.e., treatment-based changes in the microflora composition and cell density do occur. Literally, one can divide them into two principals:

(a) **Antimicrobial** manipulation, aiming at a significant reduction of the numbers of all skin microorganisms.
(b) Manipulation aiming at an increase of the number of selected, potentially beneficial microorganisms using special nutrients (**prebiotics**) or even living microorganisms (**probiotics**) (Fig. 5.3).

The overall beneficial and often life-saving effect of **antibiotics** against microbial infections is undisputed. Nevertheless, in particular studies on the human intestinal tract, such as (Dethlefsen and Relman 2011), have shown that the use of broad-spectrum antibiotics can have a pervasive and long-lasting effect on structure and (potentially the) function of this microbial ecosystem, and thereby uncouple many of the mutualistic host-microbe relationships that have been unraveled in the recent years (Modi et al. 2014). Although similar intervention studies with antibiotics are – to our knowledge – still missing, the same probably also applies to the human skin and the equilibrium of its microbiota. Consequently, therapies that make use of the variety of beneficial host–skin microbiota interactions and an intact microbiological equilibrium are increasingly investigated (for reviews see Krutmann 2009; Scharschmidt and Fischbach 2013; Al-Ghazzewi and Tester 2014; Grice 2014).

The composition of the skin microbiota depends on several factors that can easily disturb the microbial equilibrium. It is, however, particularly important that the members of the protecting/beneficial part of the microbiota are kept in a suitable balance. One way to reach and preserve a balanced skin microbiota could be by using the **prebiotic concept**. The concept of prebiotics was introduced by Gibson and Roberfroid in 1995 (Gibson and Roberfroid 1995). They defined prebiotic actives as "non-digestible food ingredients that beneficially affect the host by selectively stimulating the growth and/or activity of one, or a limited number of, bacteria in the colon". This concept was adapted by Bockmühl in 2004 to the field of cosmetic skin products (Bockmühl 2004), by introducing the prebiotic coefficient, and subsequently used by Carolan and co-workers (2008).

An increasing amount of prebiotic products was developed in recent years. Different plant extracts were investigated on their effect on *Propionibacterium acnes*, involved in acne generation, as well as on the commensal bacterium *Staphylococcus epidermidis* (Bockmühl et al. 2006). It was possible to identify plant extracts that demonstrate inhibitory effects on *P. acnes* and stimulating effects on *S. epidermidis*. In particular, a mixture of pine and black currant was very effective, whereas black currant extract alone did not work as well. These effects were of particular interest since *Propionibacterium acnes* is seen as a major reason for the development of inflamed skin conditions (see paragraph 5.4.2). Therefore, a reduction of this bacterial species appears as a suitable cosmetic treatment for inflamed and acne prone skin.

In a human study, the efficacy of the above mentioned substances in cosmetic formulations was proven. It was observed that twice daily application of a cosmetic product containing 0.5 % of selected plant extracts of pine, black currant and ginseng to human skin for a total of 3 weeks was effective in inhibiting the growth of *P. acnes*, whereas coagulase-negative staphylococci (CNS), such as *S. epidermidis*, were not affected (Bockmühl et al. 2006). The relative abundance of *P. acnes* was reduced while the relative abundance of CNS was increased. In a clinical study (Janssen and Waldmann-Laue 2008), a cosmetic formulation containing a wash

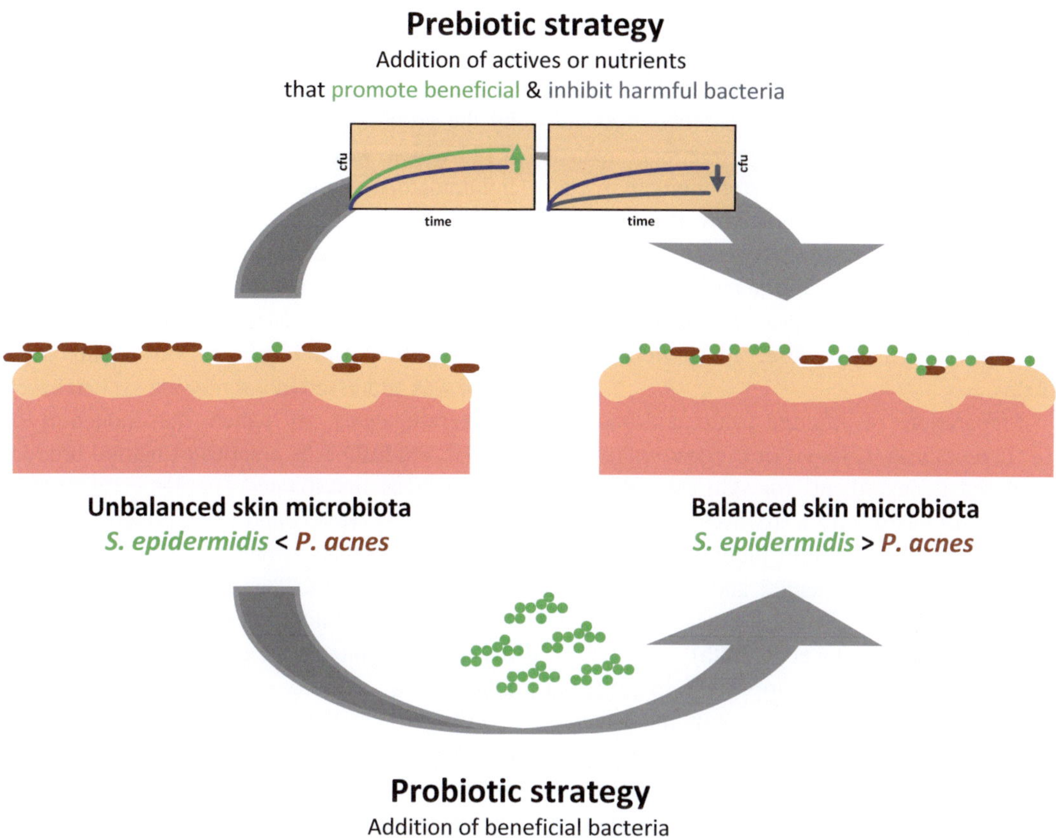


**Fig. 5.3** Pre- and probiotic concepts to rebalance the composition of the skin microbiota, e.g. in the case of impure skin (Adapted from (Krutmann 2009; Simmering and Breves 2009). *Cfu* = colony forming units, i.e. bacteria

gel, a toner and a skin fluid were tested on 30 volunteers with a mild form of impure skin. All formulas contained in total 1 % of the above mentioned prebiotic actives. A significant improvement for papules, pustules, comedones and sebum production was detected.

Another approach to balance the skin microbiota was used for people suffering from atopic dermatitis (AD) and dry skin, respectively. Although genetic factors determine the predisposition to AD, also environmental factors influence the severity of that disease. Furthermore it is known that *Staphylococcus aureus* produces a number of toxins and enzymes that often seriously worsen the state of the skin (Kozuka 2002). It was reported that *S. aureus* was detected and the microbiota was unbalanced on not only severely diseased, but also dry-type skin of patients with atopic dermatitis (Akiyama et al. 2000; Katsuyama et al. 1997; Ogawa et al. 1994; Williams et al. 1990). Katsuyama and colleagues (2005) described the use of farnesol and xylitol to balance the skin microflora of patients with atopic dermatitis. They were able to remove and prevent the adhesion of biofilm-producing *S. aureus* species on the skin.

Applying beneficial bacteria as probiotic agents directly to the skin represents another way of achieving a rebalanced microbiota situation. Ouwehand and colleagues (2003) proposed the use of propionibacteria for cosmetic products. Food grade strains were chosen as candidates, because cutaneous isolates might be correlated with skin infections. An antimicrobial activity against the skin pathogens *M. furfur*, *C. albicans* and *S. aureus*, was demonstrated, presumably due to the secretion of organic acids and the interference with the attachment of these target germs to keratin.

A rather therapeutic approach was followed in a pilot study with inactivated *Lactobacillus acidophilus*, which was used as a treatment against mild to moderate vernal keratoconjunctivitis, an allergic eye disease. After giving the probiotic containing eye-drops for 2–4 weeks, a reduction of symptoms as well as of the molecular markers ICAM-1 and TLR-4, proteins of the innate immune system, was observed (Iovieno et al. 2008).

Preparations of different lactic acid bacteria, i.e. *L. paracasei*, *L. brevis* or *L. fermentum*, were proposed as ingredients for skin care products. Cultures of lactobacilli were investigated in vitro and in vivo on the skin of individual volunteers for their probiotic potential and were found to promote *S. epidermidis* and to block *S. aureus*, *E. coli* or *M. luteus* (Lang et al. 2006).

While the **probiotic approach** in principal can be done in live form, especially in cosmetics the dosage of the beneficial microbes as inactivated biomass preparations is more widespread. This will lead to an approach of stimulating the skin's immune defence by microbial preparations or other suitable actives. The effects on the skin caused by prebiotic and probiotic actives presumably are not restricted to a direct promotion or restriction of microorganisms on the skin and oral cavity, but must be seen in the triangular relationship between the active substances, the microorganisms and the epithelial cells. Especially probiotic applications using microbial preparations probably involve the skin's innate immune system. Thus, the idea to **stimulate the skin defense** directly by actives and therapeutic agents appears very attractive (Finlay and Hancock 2004).

Donnarumma and colleagues (2004) have shown that an extract from avocado was able to influence the adherence of *M. furfur* to keratinocytes and to induce the production of human ß-defensin 2 (hBD-2). The extract mainly consisted of two rare sugars, i.e. mannoheptulose and perseitol. The observed activity of these substances might be due to a structural similarity to constituents of the cell wall of yeasts. The mode of binding to receptors and the influence on cytokine expression was also discussed.

An even more complex reaction was induced by the probiotic strain *Streptococcus salivarius* K12, which was shown to stimulate an anti-inflammatory response (Cosseau et al. 2008).

The main task in further development will be to stimulate the antimicrobial defense system without causing an overreaction that might lead to allergic reactions, severe inflammation or even sepsis (Finlay and Hancock 2004). This challenge was addressed in a screening system based on keratinocytes, in which the induction of hBD-2 and hBD-3 by a series of natural product extracts was investigated (Pernet et al. 2005). Nine extracts were found to be positive without inducing pro-inflammatory cytokines such as IL-8, IL-1α or MIP-3α. Thus, these extracts such as Arnica, Betel, Black elder and Mugwort were discussed as suitable for cosmetic or therapeutic applications.

Interestingly, the expression of the antimicrobial peptide cathelicidin LL-37 is regulated by a vitamin D3 responding element in the promoter region. The oral supplementation with vitamin D3 has been discussed as beneficial in atopic dermatitis as well as the direct topic dosage, although the latter is being hampered by skin irritation effects found in experiments with mice (Schauber and Gallo 2008).

In a more recent double-blind, placebo-controlled study, oral intake of *Lactococcus lactis* strains improved some skin properties and body characteristics in women, like skin elasticity (Kimoto-Nira et al. 2012). This and other, partly unpublished studies have led to an increasing amount of pre- and probiotic cosmetic products.

## 5.6 Outlook: Trends and Challenges

The microbiota of the human skin represents a major part of the human microbiome, which undoubtedly plays a significant role for human health and well-being. A deeper and more mechanistic (rather than descriptive) understanding of the complex interaction-network of skin microbiota, host and environment is needed as a basis for the definition of alternative diagnostic parameters as well as new preventive and therapeutic strategies for skin-related diseases and cosmetic disorders. Furthermore, the skin microbiota offers excellent opportunities for basic research on human-microbe-interactions. Future research in this field will therefore probably (continue) to focus on the following topics:

- Definition of a normal and healthy skin microbiota.
- Effect of environmental and host factors on structure (composition) and function (metabolism) of the skin microbiota.
- Interplay of bacteria, archaea, fungi, viruses and eukaryotes on the human skin.
- Interaction of the human skin microbiota with the host immune system, and vice versa.
- Mechanistic understanding of the role of skin microorganisms in skin diseases and cosmetic skin disorders in order to discriminate cause and effect.
- Interplay of the skin microbiota with the microbiota of other body compartments, e.g. the human intestinal tract.
- Potential induction of epigenetic changes in the host by members of the skin microbiota.
- Definition of microbe-based diagnostic parameters (e.g. key microbial species or microbial communities) for the early and reliable diagnosis of skin diseases.
- Development of new therapeutic strategies (prebiotic, probiotic) for skin diseases and disorders, which respect and utilize mutualistic host-microbe relationships.

## References


Akiyama H, Yamasaki O, Tada J, Arata J (2000) Adherence characteristics and susceptibility to antimicrobial agents of Staphylococcus aureus strains isolated from skin infections and atopic dermatitis. J Dermatol Sci 23:155–160

Alekseyenko AV, Perez-Perez GI, De Souza A, Strober B, Gao Z, Bihan M, Li K, Methé BA, Blaser MJ (2013) Community differentiation of the cutaneous microbiota in psoriasis. Microbiome 1:31. doi:10.1186/2049-2618-1-31

Al-Ghazzewi FH, Tester RF (2014) Impact of prebiotics and probiotics on skin health. Benefic Microbes 5:99–107. doi:10.3920/BM2013.0040

Bawdon D, Cox DS, Ashford D, James AG, Thomas GH (2015) Identification of axillary Staphylococcus sp. involved in the production of the malodorous thioalcohol 3-methyl-3-sufanylhexan-1-ol. FEMS Microbiol Lett 362:fnv111. doi:10.1093/femsle/fnv111

Bek-Thomsen M, Lomholt HB, Kilian M (2008) Acne is not associated with yet-uncultured bacteria. J Clin Microbiol 46:3355–3360. doi:10.1128/JCM.00799-08

Belkaid Y, Naik S (2013) Compartmentalized and systemic control of tissue immunity by commensals. Nat Immunol 14:646–653. doi:10.1038/ni.2604

Belkaid Y, Segre JA (2014) Dialogue between skin microbiota and immunity. Science 346:954–959. doi:10.1126/science.1260144

Bockmühl DP (2004) Präbiotika für kosmetische Anwendungen. SÖFW J 130:3–6

Bockmühl DP, Jassoy C, Nieveler S, Scholtyssek R, Wadle A, Waldmann-Laue M (2006) Prebiotic cosmetics: an alternative to antibacterial products. IFSCC Mag 9:197–200

Bojar, Holland (2002) Review: the human cutaneous microflora and factors controlling colonisation. World J Microbiol Biotechnol 18:889–903. doi:10.1023/A:1021271028979

Braff MH, Gallo RL (2006) Antimicrobial peptides: an essential component of the skin defensive barrier. Curr Top Microbiol Immunol 306:91–110

Braff MH, Bardan A, Nizet V, Gallo RL (2005) Cutaneous defense mechanisms by antimicrobial peptides. J Invest Dermatol 125:9–13. doi:10.1111/j.0022-202X.2004.23587.x

Brüggemann H (2005) Insights in the pathogenic potential of Propionibacterium acnes from its complete genome. Semin Cutan Med Surg 24:67–72. doi:10.1016/j.sder.2005.03.001

Brüggemann H, Henne A, Hoster F, Liesegang H, Wiezer A, Strittmatter A, Hujer S, Dürre P, Gottschalk G (2004) The complete genome sequence of Propionibacterium acnes, a commensal of human skin. Science 305:671–673. doi:10.1126/science.1100330

Callewaert C, Hutapea P, Van de Wiele T, Boon N (2014) Deodorants and antiperspirants affect the axillary bacterial community. Arch Dermatol Res. doi:10.1007/s00403-014-1487-1

Canesso MCC, Vieira AT, Castro TBR, Schirmer BGA, Cisalpino D, Martins FS, Rachid MA, Nicoli JR, Teixeira MM, Barcelos LS (2014) Skin wound healing is accelerated and scarless in the absence of commensal microbiota. J Immunol 193:5171–5180. doi:10.4049/jimmunol.1400625

Capone KA, Dowd SE, Stamatas GN, Nikolovski J (2011) Diversity of the human skin microbiome early in life.

J Investig Dermatol 131:2026–2032. doi:10.1038/jid.2011.168
Carolan H, Watkins S, Bradshaw D (2008) The prebiotic concept-a novel aproach for skin health. Euro Cosmet 7(8):22–27
Christensen GJM, Brüggemann H (2014) Bacterial skin commensals and their role as host guardians. Benefic Microbes 5:201–215. doi:10.3920/BM2012.0062
Clemente JC, Pehrsson EC, Blaser MJ, Sandhu K, Gao Z, Wang B, Magris M, Hidalgo G, Contreras M, Noya-Alarcón Ó, Lander O, McDonald J, Cox M, Walter J, Oh PL, Ruiz JF, Rodriguez S, Shen N, Song SJ, Metcalf J, Knight R, Dantas G, Dominguez-Bello MG (2015) The microbiome of uncontacted Amerindians. Sci Adv 1:e1500183. doi:10.1126/sciadv.1500183
Cogen AL, Yamasaki K, Sanchez KM, Dorschner RA, Lai Y, MacLeod DT, Torpey JW, Otto M, Nizet V, Kim JE, Gallo RL (2010) Selective antimicrobial action is provided by phenol-soluble modulins derived from Staphylococcus epidermidis, a normal resident of the skin. J Invest Dermatol 130:192–200. doi:10.1038/jid.2009.243
Cosseau C, Devine DA, Dullaghan E, Gardy JL, Chikatamarla A, Gellatly S, Yu LL, Pistolic J, Falsafi R, Tagg J, Hancock REW (2008) The commensal Streptococcus salivarius K12 downregulates the innate immune responses of human epithelial cells and promotes host-microbe homeostasis. Infect Immun 76:4163–4175. doi:10.1128/IAI.00188-08
Dethlefsen L, Relman DA (2011) Incomplete recovery and individualized responses of the human distal gut microbiota to repeated antibiotic perturbation. Proc Natl Acad Sci U S A 108(Suppl 1):4554–4561. doi:10.1073/pnas.1000087107
Dominguez-Bello MG, Costello EK, Contreras M, Magris M, Hidalgo G, Fierer N, Knight R (2010) Delivery mode shapes the acquisition and structure of the initial microbiota across multiple body habitats in newborns. Proc Natl Acad Sci 107:11971–11975. doi:10.1073/pnas.1002601107
Donnarumma G, Paoletti I, Buommino E, Orlando M, Tufano MA, Baroni A (2004) Malassezia furfur induces the expression of beta-defensin-2 in human keratinocytes in a protein kinase C-dependent manner. Arch Dermatol Res 295:474–481. doi:10.1007/s00403-003-0445-0
Edmonds-Wilson SL, Nurinova NI, Zapka CA, Fierer N, Wilson M (2015) Review of human hand microbiome research. J Dermatol Sci. doi:10.1016/j.jdermsci.2015.07.006
Egert M, Schmidt I, Höhne H-M, Lachnit T, Schmitz RA, Breves R (2011) rRNA-based profiling of bacteria in the axilla of healthy males suggests right-left asymmetry in bacterial activity. FEMS Microbiol Ecol 77:146–153. doi:10.1111/j.1574-6941.2011.01097.x
Egert M, Höhne H-M, Weber T, Simmering R, Banowski B, Breves R (2013) Identification of compounds inhibiting the C-S lyase activity of a cell extract from a Staphylococcus sp. isolated from human skin. Lett Appl Microbiol 57:534–539. doi:10.1111/lam.12146
Egert M, Simmering R, Banowski B, Breves R (2014) In Deo veritas—Entstehung und Verhinderung humanen Körpergeruchs. BIOspektrum 20:497–499. doi:10.1007/s12268-014-0469-3
Fierer N, Hamady M, Lauber CL, Knight R (2008) The influence of sex, handedness, and washing on the diversity of hand surface bacteria. Proc Natl Acad Sci 105:17994–17999. doi:10.1073/pnas.0807920105
Findley K, Oh J, Yang J, Conlan S, Deming C, Meyer JA, Schoenfeld D, Nomicos E, Park M, Becker J, Benjamin B, Blakesley R, Bouffard G, Brooks S, Coleman H, Dekhtyar M, Gregory M, Guan X, Gupta J, Han J, Hargrove A, Ho S, Johnson T, Legaspi R, Lovett S, Maduro Q, Masiello C, Maskeri B, McDowell J, Montemayor C, Mullikin J, Park M, Riebow N, Schandler K, Schmidt B, Sison C, Stantripop M, Thomas J, Thomas P, Vemulapalli M, Young A, Kong HH, Segre JA (2013) Topographic diversity of fungal and bacterial communities in human skin. Nature 498:367–370. doi:10.1038/nature12171
Finlay BB, Hancock REW (2004) Can innate immunity be enhanced to treat microbial infections? Nat Rev Microbiol 2:497–504. doi:10.1038/nrmicro908
Forney L, Zhou X, Brown C (2004) Molecular microbial ecology: land of the one-eyed king. Curr Opin Microbiol 7:210–220. doi:10.1016/j.mib.2004.04.015
Foulongne V, Sauvage V, Hebert C, Dereure O, Cheval J, Gouilh MA, Pariente K, Segondy M, Burguière A, Manuguerra J-C, Caro V, Eloit M (2012) Human skin microbiota: high diversity of DNA viruses identified on the human skin by high throughput sequencing. PLoS ONE 7:e38499. doi:10.1371/journal.pone.0038499
Fredrich E, Barzantny H, Brune I, Tauch A (2013) Daily battle against body odor: towards the activity of the axillary microbiota. Trends Microbiol 21:305–312. doi:10.1016/j.tim.2013.03.002
Frohm M, Agerberth B, Ahangari G, Stâhle-Bäckdahl M, Lidén S, Wigzell H, Gudmundsson GH (1997) The expression of the gene coding for the antibacterial peptide LL-37 is induced in human keratinocytes during inflammatory disorders. J Biol Chem 272:15258–15263
Fujii T, Shinozaki J, Kajiura T, Iwasaki K, Fudou R (2014) A newly discovered Anaerococcus strain responsible for axillary odor and a new axillary odor inhibitor, pentagalloyl glucose. FEMS Microbiol Ecol 89:198–207. doi:10.1111/1574-6941.12347
Gao Z, Tseng C, Pei Z, Blaser MJ (2007) Molecular analysis of human forearm superficial skin bacterial biota. Proc Natl Acad Sci U S A 104:2927–2932. doi:10.1073/pnas.0607077104
Gao Z, Tseng C, Strober BE, Pei Z, Blaser MJ (2008) Substantial alterations of the cutaneous bacterial biota in psoriatic lesions. PLoS ONE 3:e2719. doi:10.1371/journal.pone.0002719

García JR, Krause A, Schulz S, Rodríguez-Jiménez FJ, Klüver E, Adermann K, Forssmann U, Frimpong-Boateng A, Bals R, Forssmann WG (2001) Human beta-defensin 4: a novel inducible peptide with a specific salt-sensitive spectrum of antimicrobial activity. FASEB J 15:1819–1821

Giacomoni PU, Mammone T, Teri M (2009) Gender-linked differences in human skin. J Dermatol Sci 55:144–149. doi:10.1016/j.jdermsci.2009.06.001

Gibson GR, Roberfroid MB (1995) Dietary modulation of the human colonic microbiota: introducing the concept of prebiotics. J Nutr 125:1401–1412

Grice EA (2014) The skin microbiome: potential for novel diagnostic and therapeutic approaches to cutaneous disease. Semin Cutan Med Surg 33:98–103

Grice EA, Segre JA (2011) The skin microbiome. Nat Rev Microbiol 9:244–253. doi:10.1038/nrmicro2537

Grice EA, Kong HH, Conlan S, Deming CB, Davis J, Young AC, Comparative Sequencing Program NISC, Bouffard GG, Blakesley RW, Murray PR, Green ED, Turner ML, Segre JA (2009) Topographical and temporal diversity of the human skin microbiome. Science 324:1190–1192. doi:10.1126/science.1171700

Hannigan GD, Hodkinson BP, McGinnis K, Tyldsley AS, Anari JB, Horan AD, Grice EA, Mehta S (2014) Culture-independent pilot study of microbiota colonizing open fractures and association with severity, mechanism, location, and complication from presentation to early outpatient follow-up. J Orthop Res 32:597–605. doi:10.1002/jor.22578

Harder J, Schroder J-M (2002) RNase 7, a novel innate immune defense antimicrobial protein of healthy human skin. J Biol Chem 277:46779–46784. doi:10.1074/jbc.M207587200

Harder J, Bartels J, Christophers E, Schroder JM (2001) Isolation and characterization of human beta -defensin-3, a novel human inducible peptide antibiotic. J Biol Chem 276:5707–5713. doi:10.1074/jbc.M008557200

Harris RN, Brucker RM, Walke JB, Becker MH, Schwantes CR, Flaherty DC, Lam BA, Woodhams DC, Briggs CJ, Vredenburg VT, Minbiole KPC (2009) Skin microbes on frogs prevent morbidity and mortality caused by a lethal skin fungus. ISME J 3:818–824. doi:10.1038/ismej.2009.27

Horz H-P (2015) Archaeal lineages within the human microbiome: absent, rare or elusive? Life (Basel) 5:1333–1345. doi:10.3390/life5021333

Horz H-P, Conrads G (2010) The discussion goes on: what is the role of euryarchaeota in humans? Archaea 2010:1–8. doi:10.1155/2010/967271

Hulcr J, Latimer AM, Henley JB, Rountree NR, Fierer N, Lucky A, Lowman MD, Dunn RR (2012) A jungle in there: bacteria in belly buttons are highly diverse, but predictable. PLoS ONE 7:e47712. doi:10.1371/journal.pone.0047712

Iovieno A, Lambiase A, Sacchetti M, Stampachiacchiere B, Micera A, Bonini S (2008) Preliminary evidence of the efficacy of probiotic eye-drop treatment in patients with vernal keratoconjunctivitis. Graefes Arch Clin Exp Ophthalmol 246:435–441. doi:10.1007/s00417-007-0682-6

Iwase T, Uehara Y, Shinji H, Tajima A, Seo H, Takada K, Agata T, Mizunoe Y (2010) Staphylococcus epidermidis Esp inhibits Staphylococcus aureus biofilm formation and nasal colonization. Nature 465:346–349. doi:10.1038/nature09074

James AG, Austin CJ, Cox DS, Taylor D, Calvert R (2013) Microbiological and biochemical origins of human axillary odour. FEMS Microbiol Ecol 83:527–540. doi:10.1111/1574-6941.12054

Janssen F, Waldmann-Laue M (2008) Efficacy of a prebiotic product combination against skin impurities. Presented at the IFSCC Conference, Barcelona

Julian TR, Leckie JO, Boehm AB (2010) Virus transfer between fingerpads and fomites: virus transfer between fingerpads and fomites. J Appl Microbiol 109:1868–1874. doi:10.1111/j.1365-2672.2010.04814.x

Katsuyama M, Wachi Y, Ikezawa Z, Kitamura K, Suga C, Ohnuma S (1997) Correlation between the population of Staphylococcus aureus on the skin and severity of a score of dry type atopic dermatitis conditions. Nippon Hifuka Gakkai Zasshi 107:1103–1111

Katsuyama M, Masako K, Kobayashi Y, Yusuke K, Ichikawa H, Hideyuki I, Mizuno A, Atsuko M, Miyachi Y, Yoshiki M, Matsunaga K, Kayoko M, Kawashima M, Makoto K (2005) A novel method to control the balance of skin microflora Part 2. A study to assess the effect of a cream containing farnesol and xylitol on atopic dry skin. J Dermatol Sci 38:207–213. doi:10.1016/j.jdermsci.2005.01.003

Kelly DP, Wood AP (2010) Skin microbiology, body odor, and methylotrophic bacteria. In: Timmis KN (ed) Handbook of hydrocarbon and lipid microbiology., Biomedical and life sciences. Springer, Heidelberg, pp 3203–3213

Kimoto-Nira H, Aoki R, Sasaki K, Suzuki C, Mizumachi K (2012) Oral intake of heat-killed cells of Lactococcus lactis strain H61 promotes skin health in women. J Nutr Sci 1:e18. doi:10.1017/jns.2012.22

Kong HH (2011) Skin microbiome: genomics-based insights into the diversity and role of skin microbes. Trends Mol Med 17:320–328. doi:10.1016/j.molmed.2011.01.013

Kong HH, Segre JA (2012) Skin microbiome: looking back to move forward. J Investig Dermatol 132:933–939. doi:10.1038/jid.2011.417

Kong HH, Oh J, Deming C, Conlan S, Grice EA, Beatson MA, Nomicos E, Polley EC, Komarow HD, Comparative Sequence Program NISC, Murray PR, Turner ML, Segre JA (2012) Temporal shifts in the skin microbiome associated with disease flares and treatment in children with atopic dermatitis. Genome Res 22:850–859. doi:10.1101/gr.131029.111

Kozuka T (2002) Patch testing to exclude allergic contact dermatitis caused by povidone-iodine. Dermatol (Basel) 204(Suppl 1):96–98, doi:57734

Krutmann J (2009) Pre- and probiotics for human skin. J Dermatol Sci 54:1–5. doi:10.1016/j.jdermsci.2009.01.002

Lai Y, Di Nardo A, Nakatsuji T, Leichtle A, Yang Y, Cogen AL, Wu Z-R, Hooper LV, Schmidt RR, von Aulock S, Radek KA, Huang C-M, Ryan AF, Gallo RL (2009) Commensal bacteria regulate Toll-like receptor 3-dependent inflammation after skin injury. Nat Med 15:1377–1382. doi:10.1038/nm.2062

Lai Y, Cogen AL, Radek KA, Park HJ, Macleod DT, Leichtle A, Ryan AF, Di Nardo A, Gallo RL (2010) Activation of TLR2 by a small molecule produced by Staphylococcus epidermidis increases antimicrobial defense against bacterial skin infections. J Invest Dermatol 130:2211–2221. doi:10.1038/jid.2010.123

Lang C, Heilmann A, Veen M, Budde E, Böttner M, Reindl A, Knöll R (2006) Methods and means for protecting the skin against pathogenic bacteria. U.S. patent WO 2006/136420 A2

Mao G-Y, Yang S-L, Zheng J-H (2008) Etiology and management of axillary bromidrosis: a brief review. Int J Dermatol 47:1063–1068. doi:10.1111/j.1365-4632.2008.03735.x

McAleer MA, Irvine AD (2013) The multifunctional role of filaggrin in allergic skin disease. J Allergy Clin Immunol 131:280–291. doi:10.1016/j.jaci.2012.12.668

Meadow JF, Bateman AC, Herkert KM, O'Connor TK, Green JL (2013) Significant changes in the skin microbiome mediated by the sport of roller derby. Peer J 1:e53. doi:10.7717/peerj.53

Midorikawa K, Ouhara K, Komatsuzawa H, Kawai T, Yamada S, Fujiwara T, Yamazaki K, Sayama K, Taubman MA, Kurihara H, Hashimoto K, Sugai M (2003) Staphylococcus aureus susceptibility to innate antimicrobial peptides, beta-defensins and CAP18, expressed by human keratinocytes. Infect Immun 71:3730–3739

Min YW, Rhee P-L (2015) The role of microbiota on the gut immunology. Clin Ther 37:968–975. doi:10.1016/j.clinthera.2015.03.009

Minot S, Sinha R, Chen J, Li H, Keilbaugh SA, Wu GD, Lewis JD, Bushman FD (2011) The human gut virome: inter-individual variation and dynamic response to diet. Genome Res 21:1616–1625. doi:10.1101/gr.122705.111

Modi SR, Collins JJ, Relman DA (2014) Antibiotics and the gut microbiota. J Clin Invest 124:4212–4218. doi:10.1172/JCI72333

Naik S, Bouladoux N, Wilhelm C, Molloy MJ, Salcedo R, Kastenmuller W, Deming C, Quinones M, Koo L, Conlan S, Spencer S, Hall JA, Dzutsev A, Kong H, Campbell DJ, Trinchieri G, Segre JA, Belkaid Y (2012) Compartmentalized control of skin immunity by resident commensals. Science 337:1115–1119. doi:10.1126/science.1225152

Nakamizo S, Egawa G, Honda T, Nakajima S, Belkaid Y, Kabashima K (2015) Commensal bacteria and cutaneous immunity. Semin Immunopathol 37:73–80. doi:10.1007/s00281-014-0452-6

Nakatsuji T, Chiang H-I, Jiang SB, Nagarajan H, Zengler K, Gallo RL (2013) The microbiome extends to subepidermal compartments of normal skin. Nat Commun 4:1431. doi:10.1038/ncomms2441

Noël F, Piérard-Franchimont C, Piérard GE, Quatresooz P (2012) Sweaty skin, background and assessments. Int J Dermatol 51:647–655. doi:10.1111/j.1365-4632.2011.05307.x

Nomura I, Gao B, Boguniewicz M, Darst MA, Travers JB, Leung DY m (2003) Distinct patterns of gene expression in the skin lesions of atopic dermatitis and psoriasis: a gene microarray analysis. J Allergy Clin Immunol 112:1195–1202. doi:10.1016/j.jaci.2003.08.049

Ogawa T, Katsuoka K, Kawano K, Nishiyama S (1994) Comparative study of staphylococcal flora on the skin surface of atopic dermatitis patients and healthy subjects. J Dermatol 21:453–460

Oh J, Freeman AF, Comparative Sequencing Program NISC, Park M, Sokolic R, Candotti F, Holland SM, Segre JA, Kong HH (2013) The altered landscape of the human skin microbiome in patients with primary immunodeficiencies. Genome Res 23:2103–2114. doi:10.1101/gr.159467.113

Ong PY, Ohtake T, Brandt C, Strickland I, Boguniewicz M, Ganz T, Gallo RL, Leung DYM (2002) Endogenous antimicrobial peptides and skin infections in atopic dermatitis. N Engl J Med 347:1151–1160. doi:10.1056/NEJMoa021481

Ouwehand AC, Båtsman A, Salminen S (2003) Probiotics for the skin: a new area of potential application? Lett Appl Microbiol 36:327–331

Park HK, Ha M-H, Park S-G, Kim MN, Kim BJ, Kim W (2012) Characterization of the fungal microbiota (mycobiome) in healthy and dandruff-afflicted human scalps. PLoS ONE 7:e32847. doi:10.1371/journal.pone.0032847

Percival SL, Emanuel C, Cutting KF, Williams DW (2012) Microbiology of the skin and the role of biofilms in infection. Int Wound J 9:14–32. doi:10.1111/j.1742-481X.2011.00836.x

Pernet I, Reymermier C, Guezennec A, Viac J, Guesnet J, Perrier E (2005) An optimized method for intensive screening of molecules that stimulate beta-defensin 2 or 3 (hBD2 or hBD3) expression in cultured normal human keratinocytes. Int J Cosmet Sci 27:161–170. doi:10.1111/j.1467-2494.2005.00262.x

Philpott MP (2003) Defensins and acne. Mol Immunol 40:457–462

Picardo M, Ottaviani M (2014) Skin microbiome and skin disease: the example of rosacea. J Clin Gastroenterol 48(Suppl 1):S85–S86. doi:10.1097/MCG.0000000000000241

Probst AJ, Auerbach AK, Moissl-Eichinger C (2013) Archaea on human skin. PLoS ONE 8:e65388. doi:10.1371/journal.pone.0065388

Redel H, Gao Z, Li H, Alekseyenko AV, Zhou Y, Perez-Perez GI, Weinstock G, Sodergren E, Blaser MJ (2013) Quantitation and composition of cutaneous microbiota in diabetic and nondiabetic men. J Infect Dis 207:1105–1114. doi:10.1093/infdis/jit005

Rosenthal M, Goldberg D, Aiello A, Larson E, Foxman B (2011) Skin microbiota: microbial community structure and its potential association with health and disease. Infect Genet Evol 11:839–848. doi:10.1016/j.meegid.2011.03.022

Roth RR, James WD (1988) Microbial ecology of the skin. Annu Rev Microbiol 42:441–464. doi:10.1146/annurev.mi.42.100188.002301

Sanford JA, Gallo RL (2013) Functions of the skin microbiota in health and disease. Semin Immunol 25:370–377. doi:10.1016/j.smim.2013.09.005

SanMiguel A, Grice EA (2015) Interactions between host factors and the skin microbiome. Cell Mol Life Sci 72:1499–1515. doi:10.1007/s00018-014-1812-z

Sato K, Kang WH, Saga K, Sato KT (1989) Biology of sweat glands and their disorders. I. Normal sweat gland function. J Am Acad Dermatol 20:537–563

Scharschmidt TC, Fischbach MA (2013) What lives on our skin: ecology. Genomics and therapeutic opportunities of the skin microbiome. Drug Discov Today Dis Mech 10:e83–e89. doi:10.1016/j.ddmec.2012.12.003

Schauber J, Gallo RL (2008) Antimicrobial peptides and the skin immune defense system. J Allergy Clin Immunol 122:261–266. doi:10.1016/j.jaci.2008.03.027

Schommer NN, Gallo RL (2013) Structure and function of the human skin microbiome. Trends Microbiol 21:660–668. doi:10.1016/j.tim.2013.10.001

Seite S, Flores GE, Henley JB, Martin R, Zelenkova H, Aguilar L, Fierer N (2014) Microbiome of affected and unaffected skin of patients with atopic dermatitis before and after emollient treatment. J Drugs Dermatol 13:1365–1372

Simmering R, Breves R (2009) Pre- and probiotic cosmetics. Hautarzt 60:809–814. doi:10.1007/s00105-009-1759-4

Tomic-Canic M, Perez-Perez GI, Blumenberg M (2014) Cutaneous microbiome studies in the times of affordable sequencing. J Dermatol Sci 75:82–87. doi:10.1016/j.jdermsci.2014.05.001

Troccaz M, Gaïa N, Beccucci S, Schrenzel J, Cayeux I, Starkenmann C, Lazarevic V (2015) Mapping axillary microbiota responsible for body odours using a culture-independent approach. Microbiome 3:3. doi:10.1186/s40168-014-0064-3

Urmacher C (1990) Histology of normal skin. Am J Surg Pathol 14:671–686

Verhulst NO, Qiu YT, Beijleveld H, Maliepaard C, Knights D, Schulz S, Berg-Lyons D, Lauber CL, Verduijn W, Haasnoot GW, Mumm R, Bouwmeester HJ, Claas FHJ, Dicke M, van Loon JJA, Takken W, Knight R, Smallegange RC (2011) Composition of human skin microbiota affects attractiveness to malaria mosquitoes. PLoS ONE 6:e28991. doi:10.1371/journal.pone.0028991

Wilke K, Martin A, Terstegen L, Biel SS (2007) A short history of sweat gland biology. Int J Cosmet Sci 29:169–179. doi:10.1111/j.1467-2494.2007.00387.x

Williams RE, Gibson AG, Aitchison TC, Lever R, Mackie RM (1990) Assessment of a contact-plate sampling technique and subsequent quantitative bacterial studies in atopic dermatitis. Br J Dermatol 123:493–501

Wilson M (2008) Bacteriology of humans an ecological perspective. Blackwell Pub, Malden

Zeeuwen PLJM, Kleerebezem M, Timmerman HM, Schalkwijk J (2013) Microbiome and skin diseases. Curr Opin Allergy Clin Immunol 13:514–520. doi:10.1097/ACI.0b013e328364ebeb

# Vaginal Microbiota

6

Werner Mendling

**Abstract**

The knowledge about the normal and abnormal vaginal microbiome has changed over the last years. Culturing techniques are not suitable any more for determination of a normal or abnormal vaginal microbiota. Non culture-based modern technologies revealed a complex and dynamic system mainly dominated by lactobacilli.

The normal and the abnormal vaginal microbiota are complex ecosystems of more than 200 bacterial species influenced by genes, ethnic background and environmental and behavioral factors. Several species of lactobacilli per individuum dominate the healthy vagina. They support a defense system together with antibacterial substances, cytokines, defensins and others against dysbiosis, infections and care for an normal pregnancy without preterm birth.

The numbers of *Lactobacillus (L.) iners* increase in the case of dysbiosis.

Bacterial vaginosis (BV) – associated bacteria (BVAB), *Atopobium vaginae* and *Clostridiales* and one or two of four *Gardnerella vaginalis* – strains develop in different mixtures and numbers polymicrobial biofilms on the vaginal epithelium, which are not dissolved by antibiotic therapies according to guidelines and, thus, provoke recurrences.

Aerobic vaginitis seems to be an immunological disorder of the vagina with influence on the microbiota, which is here dominated by aerobic bacteria (*Streptococcus agalactiae, Escherichia coli*). Their role in AV is unknown.

W. Mendling (✉)
Deutsches Zentrum für Infektionen in Gynäkologie und Geburtshilfe, St. Anna Hospital, Vogelsangstrasse, 106, 42109 Wuppertal, Germany
e-mail: w.mendling@t-online.de;
http://www.werner-mendling.de



A. Schwiertz (ed.), *Microbiota of the Human Body*, Advances in Experimental Medicine and Biology 902, DOI 10.1007/978-3-319-31248-4_6

Vaginal or oral application of lactobacilli is obviously able to improve therapeutic results of BV and dysbiosis.

**Keywords**
Vaginal microbiota • Dysbiosis • Bacterial vaginosis • Aerobic vaginitis • Lactobacilli • Probiotics

## 6.1 A Historic Perpective

Albert Döderlein (1860–1941) was the first one to discover the importance of lactic acid producing bacteria in the vagina (Döderlein 1892). Krönig (1895), a co-worker of Döderlein, described lactobacilli as anaerobic and curved rods, which were later cultured by Curtis (1913) and named *Mobiluncus curtisii* (Spiegel and Roberts 1984). Finally, Stanley Thomas (1928) coined the term *Lactobacillus acidophilus*. In the 1980s Lauer, Helming and Kandler were able to distinguised several Lactobacillus species previously termed L. acidophilus by DNA-DNA hybridisation.

At the beginning of the last century first attempts were made to grade the vaginal microbiota. Manu af Heurlin (1914) characterised the vaginal microbiota of children, pregnant and not pregnant women and women of old age and tried to establish grades of healthines. Robert Schröder (1921) was the first one to define three bacteriologically different vaginal microbiota types termed "Reinheitsgrade" (grades of purity). Otto Jirovec (Jirovec et al. 1948) distinguished between six vaginal microbiota types (normal, abnormal, abnormal with many leucocytes, gonorrhea, trichomoniasis, candidosis).

In 1955 Herman Gardner and Charles Dukes (1955) described *Haemophilus vaginalis*, later renamed as *Gardnerella vaginalis* (Greenwood and Pickett 1980) as the main causative for bacterial vaginosis (BV), the most common disturbance of the vaginal microbiota and considered it as a sexually transmitted disease. Furthermore, they emphasised the importance of microscopy and defined "clue cells" as diagnostic marker.

By the end of the last century the hitherto valid definition of bacterial vaginosis was described as "a replacement of lactobacilli by characteristic groups of bacteria accompanied by changed properties of the vaginal fluid" (Mardh et al. 1984). With the advent of molecular and genetic technologies we had to reconsider our view and definition of a normal vaginal microbiota. New bacteria were discovered and the concepts of a bacterial biofilm and of vaginal microbiota types was introduced.

## 6.2 Normal Vaginal Microbiota

The outer and inner surfaces of a child born by vaginal delivery are primarily colonised by the vaginal microbiota of the mother. Futher colonisers are aquired from the skin and mouth microbiota of the mother. In the last years it became evident that mothermilk harbours a unique microbiota, mainly dominated by Lactobacilli, which is transfered to the suckling child (Martin et al. 2003). Before the menarche the vagina microbiota is a unsteady mix of skin and gut microbes, which may harbor some lactobacilli (Fettweis et al. 2012). The environmental conditions for lactobacilli become improved by estrogens and progesterone with the start and during the reproductive phase of women. Estrogens support the proliferation of the vaginal epithelium and the development of intraepithelial glycogen, while progesterone supports the cytolysis of epithelial cells, which release glycogen. Lactobacilli and other bacteria are able to metabolise this glycogen to glucose and maltose and further to lactic acid. This leads to a vaginal pH of 3.8–4.4 which is defined as normal.

Until now more than 120 *lactobacillus* species have been described (de Vos et al. 2012). Within the vaginas of women of reproductive age more than ten different species can be found. However, a single women is usually dominated by one or two species, of which the most frequent are *L. crispatus*, *L. gasseri*, *L. jensenii and L. iners* (Vasquez et al. 2002). Several of these lactobacilli are able to produce bacteriocines, biosurfactants and coaggregating molecules to inhibit the adhesion of pathogens (Reid 2001). Another property of lactobacilli found within the vagina is the ability to produce hydrogen peroxid ($H_2O_2$). Lactobacilli are by definition strict anaerobic bacteria. However, they are often found in niches enriched with oxygen. To detoxify the otherwise toxic oxygen, several but not all lactobacilli are able to produce $H_2O_2$. The presence of $H_2O_2$ producing lactobacilli is negatively associated with the formation of BV (Eschenbach et al. 1989).

### 6.2.1 The Normal Vaginal Microbiota: A Mixture of Many Bacteria in a Balance

Currently, the dogma, that a healthy vaginal microbiota is dominated by lactobacill is faltering as by genomic sequencing over 250 species of bacteria have been identified in the vagina (Li et al. 2012). Besides Lactobacilli many other bacteria can be found in the normal or abnormal vaginal microbiota, such as *Actinomyces*, *Aerococcus*, *Allisonella*, *Alloscardovia*, *Anaerococcus*, *Arcanobacterium*, *Atopobium*, *Bacteroides*, *Balneimonas*, *Bifidobacterium*, *Blastococcus*, *Blautia*, *Bulleidia*, *Campylobacter*, *Citrobacter*, *Coriobacteriacea*, *Corynebacterium*, *Enterobacter*, *Escherichia*, *Facklamia*, *Faecalibacterium*, *Finegoldia*, *Gardnerella*, *Gemella*, *Haemophilus*, *Lachnospiracea*, *Massilia*, *Megasphera*, *Mobiluncus*, *Mollicutes*, *Moryella*, *Olsinella*, *Parvimonas*, *Peptinophilus*, *Peptostreptococcus*, *Prevotella*, *Porphyromonas*, *Proteobacteria*, *Providencia*, *Rhizobialis*, *Ruminococcaceae*, *Salmonella*, *Shigella*, *Shuttleworthia*, *Sneathia*, *Solobacterium*, *Staphylococcus*, *Streptococcus*, *Veillonella*, *Ureaplasma*, and many lactobacilli species (Gajer et al. 2012).

Within the human microbiome project, the vaginal microbiome project investigated the relationship between the vaginal microbiota and various physiological and infectious conditions (Fettweis et al. 2012). Various "vagitypes" have been identified of which many are dominated by a single bacterial taxon, others by a broad spectrum of different bacteria. Interestingly the ethnic background of women has an influence on the vaginal microbiota, as white/caucasian women are dominated by *L. iners*, asian women by *L. crispatus* and black and hispanic women by *L. jenseni. However, a* significant group of women harbored no lactobacilli in the vagina (Ravel et al. 2011, Hickley et al. 2012).

Jespers et al. (2012) identified in Antverpen/Belgium similarly three types of vaginal microbiota in healthy premenopausal women and in women at risk for BV of a STD clinic.

One group of women was dominated by *L. crispatus*, *L. iners*, *L. jensenii* and *L. vaginalis* with lower counts (<30 %) of *L. gasseri* and *Atopobium vaginae*. A second group harbored preferentially *L. gasseri* and *L. vaginalis*, but less *L. jensenii*, *L. iners* or *L. crispatus*. The third group was dominated by *L. gasseri*, *A. vaginae* and *L. iners*. Whithin the third group were mainly african and asian women. These flora types underly dynamic variations during the menstrual cycle and are influenced by external circumstances, for instance sexual behavior. But they seem to be in a rather stable balance, and a healthy vaginal system can obviously be strong enough to correct disturbences from outside, as Gajer et al. (2012) demonstrated in a longitudinal study. Women were grouped according to Ravel's et al. (2011) "community state types" and vaginal swaps were taken for 16 successive weeks. Furthermore, menstruation, tampon use, vaginal, anal or oral sex, sex toys, digital penetration and lubricants were documented. It was evident, that the vaginal microbiota of several women became heavily disturbed by some of these actions, however other microbiotas showed no disturbences despite very frequent manipulations. Once again, black women were significantly different in their "community state types".

The vaginal flora is influenced by the anal and the oral flora. Petricevic et al. (2012) found in around 80 % of 30 pregnant women and in 40 % of 30 postmenopausal women one or more *Lactobacillus ssp.* in the vagina and in the rectum, and they were in 80 % resp. 40 % of the same identity. These women were also in 50 % (pregnant) and in 30 % (postmenopausal) colonised by one or more *Lactobacillus ssp.* in their mouth. A healthy vaginal, balanced microbiota protects not only against ascending infections or HIV acquisition, but also against prematurity (Hoyme and Hübner 2010; Donders et al. 2011; Lamont et al. 2011; Martin 2012; Mendling et al. 2013).

On the other hand, too many vaginal lactobacilli (Cibley and Cibley 1991) or abnormal long lactobacilli (Horowitz et al. 1994) can cause vestibular pruritus, itching and dysuria. This "cytolytic vaginosis" or "lactobacillosis" can be misdiagnosed clinically as candidosis (Demirezen 2003).

### 6.2.2 Gene Polymorphisms and Vaginal Immunity

The vaginal microbiota is not only influenced by the ethnic background, but also by gene polymorphisms: the individual capacity to produce low or high levels of anti- or pro-microbial factors influences the composition of the vaginal microbiota. Polymorphisms in the interleukin-1 receptor antagonist gene or the Toll-like receptor (TLR) 4, which acts in the innate recognition of Gram-negative bacteria, influence the quantity of vaginal bacteria (Rodriguez et al. 1999) and can influence individual susceptibility to pregnancy complications (Genc and Onderdonk 2011). Such polymorphisms vary between different racial groups and may be associated with the different ecosystems between different populations (Linhares et al. 2010). Interestingly, periodontal disease and BV are influenced by gene polymprphisms and are both associated with preterm birth (Sanu and Lamont 2011).

The innate immune system of the vagina is represented by soluble factors like mannose-binding lectin (MBL), defensins, secretory leucocyte protease inhibitor, nitric oxid, and membrane-associated factors, the TLR (11 TLR have been identified) and phagocytes. Different TLR recognize lipoproteins and peptidoglucan in the surface of Gram-positive bacteria, the lipopolysaccharid of Gram-negative bacteria, flagellins, and others (Linhares et al. 2010; Mirmonsef et al. 2011). Vaginal cells release defensins with a non-specific antimicrobial activity. The production of special defensins is stimulated by estrogens and inhibited by progesterone. Bacterial vaginosis in pregnant women was associated with lower vaginal concentrations of defensin 3 (Mitchell et al. 2013). Women with MBL deficiency due to a polymorphism are more susceptible to recurrent *Candida albicans* vaginitis (Babula et al. 2003).

Toll-like receptor ligands and fatty acids, which are produced by many vaginal bacteria, have dramatic effects on the vaginal immune function: the anaerobes of BV produce bad smelling amines (putrescin, cadaverin and others), succinate, sialidases, and immunomodulatory substances such as lipopolysaccharides, lipoteichoid acids and peptidoglycans with many influences on cytokines and other immune responses (Mirmonsef et al. 2011).

## 6.3 Abnormal Vaginal Flora

A disturbed vaginal microbiota may be the cause for various diseases. However, within this chapter only two will be discussed, as only they are directly connected to a dysbiotic vaginal microbiota.

### 6.3.1 Bacterial Vaginosis (BV)

Gardner and Dukes (1955) named the vaginal disorder *Haemophilus vaginalis* vaginitis and described "clue cells". It was later characterized as bacterial vaginosis and is defined by a replacement of lactobacilli with characteristic groups of bacteria accompanied by changed properties of the vaginal fluid (Weström et al. 1984).

The first diagnostic criteria for BV were published by Amsel et al. (1983): grey-white milky discharche, pH>4.5, bad "fishy" smell, especially if 10 % KOH solution is added, and at least 20 % "clue cells". Later, Eschenbach et al. (1989) determined the lack of $H_2O_2$-producing lactobacilli, an overgrowth of G. vaginalis and anaerobic Gram – negative rods and anaerobic Gram – positive cocci as essential factors for the presence of BV. To improve the diagnostic analyis, Nugent et al. (1991) poposed a score (Nugent score): 0–3 = normal, 4–6 = intermediate, 7–10 = BV. It is solely based on Gram-staining criteria. However, it has been reported, that roughly 20 % of pregnant women in Germany have BV by definition, but not all suffered from symptoms (Mendling et al. 2013).

The development of BV was long associated with the presence of G. vaginalis. Currently, four different *G. vaginalis* strains have been described, of which only two produce the BV marker sialidase and only one predominated in women with BV (Jayaprakash et al. 2012). Hence, the existance of *G. vaginalis* in the vagina is no precondition of BV. In the last years it became evident, that no single strain alone is the cause of BV. Recently, BV associated bacteria (BVAB) 1, 2 and 3 have been described. Nearly all of these bacteria are unknown in clinical practice. Women, who harbor BVAB, especially *G. vaginalis* and *Leptotrichia/Snethia* or *Megasphera* in higher concentrations, develop significantly more BV (p=0.001) (Marrazzo et al. 2012; Hillier et al. 2010). Additionally, Fredricks et al. (2005)˙ demonstrated that the presence of *L. iners* is strongly associated with BV. *L. iners*, which belongs to the normal microbiota, but seems to be a "poisened apple in the basket", because its presence is strongly connected with a shift of normal to abnormal microbiota. On the other hand, Women, who harbor *L. crispatus* are significantly less at risk to develop BV than others (p=0.02).

BV is influenced by environmental and genetic factors. Thus, gene polymorphisms influence the occurrence of *G. vaginalis* and *A. vaginae* (Verstraelen et al. 2009). Furthermore, decreasing estrogen levels influence the number and diversity of vaginal lactobacilli and are in some women a risk factor for urogenital infections.

Sexual practices, especially receptive oral sex and digital vaginal penetration are significant risk factors for BV (which is perhaps an explanation for a higher risk of BV in lesbian women (Marrazzo et al. 2012)), and also cigarette smoking, black race, receptive anal sex before vaginal intercourse (Cherpes et al. 2008; Manhart et al. 2012). It should be kept in mind, that Gardner and Dukes (1955) could not cause BV by transferring cultivated *G. vaginalis* from a woman with BV to a healthy woman, but if they transferred the discharge of a woman with BV to a healthy vagina, this woman got BV. Hence, not single bacteria, albeit in high numbers, is important, but a critical mixture of BVAB together with special lactobacilli, and a lack of other lactobacilli seem to play a role in the development of BV (Lamont et al. 2011). Lamont et al. (2011) discussed, "that it is whether or not the strain/species of *Lactobacillus* produces $H_2O_2$ that dictates whether BV is present or absent. However, given that $H_2O_2$ – producing *L. gasseri* are found in BV patients, albeit at lower incidence, one might also argue that in vitro production of $H_2O_2$ is only a biomarker of a protective species of *Lactobacillus*, not an active factor in limiting the growth of vaginal anaerobes."

### 6.3.2 Polymicrobial Bacterial Biofilms in BV and Sexual Transmission

A biofilm is defined as any group of microorganisms in which cells stick to each other on a surface. The first biofilm in gynecology was described in women with BV by Swidsinski et al. (2005). The epithelial cells of the vagina of healthy pre- or postmenopausal women or of children are free of bacteria. But BV is characterized by structured polymicrobial biofilms adherent to epithelial cells of the vagina. "Clue cells", which Gardner and Dukes (1955) have seen microscopically on vaginal epithelial cells of the discharge, have their origin from this biofilm – coat on the vaginal wall. The biofilm consists in its majority of *G. vaginalis* (>50 % up to 90 %) and of *A. vaginae* (10–40 %), but also of lactobacilli and other bacteria. *G. vaginalis* formes in vitro significantly stronger biofilms, if *Fusobacterium nucleatum* or *Prevotella bivia* are added (Machado et al. 2013). No BV is equal in the composition of different bacteria and no biofilm of BV is equal.

It is unknown, whether the lactobacilli found in the biofilm are *L. iners* or other species. If treated with metronidazole, it does not get disrupted and, thus, seems to be the reason for the high recurrence rates of about 30 % after 3 months and 60 % after 6 months following therapy, respectively.

In addition to be present in the vagina, the BV-typical biofilm can be found on epithelial cells in the urine of females with BV and in the urine of their partners. Sometimes it may be found in cryopreserved donor semen, in the endometrium of non-pregnant women and in tissue of missed abortion/abortion (Swidsinski et al. 2013). If men were asked to void their urine after having pulled back the preputium, no biofilm was found, which confirms the observation that male circumcision reduces the risk for ulcerations, trichomoniasis and BV (Gray et al. 2009). Circumcision is associated with a significant change in the microbiota and with a significant decrease in putative anaerobic bacteria, especially *Clostridiales* and *Prevotellaceae* (Price et al. 2010). Women with treated BV have a higher risk for recurrence, if they have intercourse with the same partner without using condoms (Marrazzo et al. 2012; Guédou et al. 2013).

### 6.3.3 Aerobic Vaginitis (AV)

In 2002 Donders et al. (2002) characterised a new type of vaginitis. It is in contrast to BV dominated by aerobic bacteria, mainly *Streptococcus agalactiae* and *Escherichia coli* and named aerobic vaginitis (AV). The patients suffer from yellow-green discharge, the vagina is red by inflammation, the pH is 5.5–6.5, many toxic leucocytes, parabasal cells and a sparse coccoid flora without lactobacilli dominate the microscopic field. High levels of interleukin – 1 beta, – 6, – 8 and leukaemia inhibiting factor in contrast to BV. Severe cases resemble to desquamative inflammatory vaginitis, which is discussed to be an early form of Lichen ruber of the vagina. AV is a higher risk factor for preterm labour and preterm birth than BV (Donders et al. 2011). Some believe, that AV is primarily an immunologic disorder with secondary abnormal microbiota, or a dermatological disease in the vagina (Edwards 2010). Women with AV are at risk for low grade cervical squamous intraepithelial cell lesions (Jahic et al. 2013). About 5 % of women in reproductive age are suffering from AV, but some diagnosed it in a much higher frequence of 23 % (Fan et al. 2013). But partner treatment is without benefit for the woman with BV.

## 6.4 Prophylaxis and Therapy with Probiotics

Probiotics are microorganisms that provide a health benefit to the host. They act in the gastrointestinal tract and influence in various ways the immune system (Sherman et al. 2009). Although lactobacilli are in use for prophylaxis or treatment of vaginal discharge since decades, probiotic research devoleped rapidly over the last 30 years within the field of gynaecology (Spurbeck and Arvidson 2011; Reid 2012). One of the first clinical studies proposed the daily oral application of about 250 g Yoghurt containing *L. acidophilus* for 6 months to women with recurrent candida vulvovaginitis. The mean rate of recurrences in the control arm was 2.5 versus 0.38 in the yoghurt arm (p=0.001) (Hilton et al. 1992). Since then, several species have been tested in various studies. *L. rhamnosus Lcr 35* (Coudeyras et al.2008a, b) showed increased ability to metabolise glycogen to lactic acid and in vitro growth inhibition of *G. vaginalis* and *C. albicans*. The effect was higher after the manufactering process than compared to three other *L. rhamnosus* strains (Nivoliez et al. 2012). Lcr 35 adheres to cervicovaginal cells and is an antagonist of BVAB (Coudeyras et al. 2008a, b).

The strain *L. rhamnosus* GR-1 causes significant killing of *E. coli* in vitro and is able to cause bacterial death in BV biofilms in vitro (McMillan et al. 2011). The two strains *L. rhamnosus* GR-1 and *L. reuteri* RC-14 (formerly *L. fermentum*), which are traded in a vaginal tablet, inhibited in vitro the growth of *C. albicans* and upregulated inflammatory Interleukin levels in a human vaginal epithelial cell line (Martinez et al. 2009a, b). *C. albicans* lost it's metabolic activity, showed increased expression of stress-related genes and lower expression of genes involved in fluconazole resistance (Köhler et al. 2012). Similar results were demonstrated by Sanchez et al. (2008) with a different strain, *L. rhamnosus* GG, which showed in a monolayer cell culture protection against damage by *C. albicans*, modulation of immune responses and immune conditioning of the mucosal surfaces (Schaller 2012). Probiotics, here administered as a daily probiotic drink for 6 months, can also enhance the clearance of human papillomavirus-related cervical lesions significantly against placebo (Verhoeven et al. 2013).

Clinical studies had been performed with different probiotics administered vaginally or orally. *L. crispatus* CTV-05 is one of the new probiotics in gynecology and well tolerated (Hemmerling et al. 2010). Vaginal intercourse (seminal fluid), and the presence of lactobacilli of the same species during vaginal administration inhibit the colonisation (Antonio et al. 2009). There seems to be a competition of one's own and the foreign lactobacilli.

Oral administration of lactobacilli to influence the vaginal microbiota seems to be effective. The first to demonstrate this were Hilton et al. (1992) against *Candida* vaginitis and Shalev et al. (1996) against *Candida* and/or BV. Rectal lactobacilli with vaginal tropism can colonise the vagina and vice versa. Oral application of a mixture of $10^8$ *L. fermentum* 57A, *L. plantarum* 57B and *L. gasseri* 57C daily for 60 days was able to colonise the rectum and the vagina between day 20 and 70 and decreased the vaginal pH, while the Nugent score improved (Strus et al. 2012). The oral administration of *L. rhamnosus* GR-1 and *L. reuteri* RC-14 for 30 days following treatment of BV with oral metronidazole improved the cure rate (Anukam et al. 2006). Furthermore, Bohbot and Cardot (2012) showed in a pilot study, that oral administration of *L. rhamnosus* Lcr35 is able to improve the Nugent score to normal values.

In addition to the improvement of BV symptoms the recurrences of vulvovaginal candidosis can be influenced by probiotics (Homayouni et al. 2014, Huang et al. 2013). Ehrström et al. (2010) showed improved treatment rates for women with BV and vulvovaginal candidosis by additional administration of *L. gasseri* LN40, *L. fermentum* LN99, *L. casei subsp. rhamnosus* LN113 and *P. acidilactici* LN23 for 5 days in vaginal capsules. Martinez et al. (2009a, b) improved the clinical treatment results of vulvovaginal candidosis with oral fluconazole, *L. rhamnosus* GR-1 and *L. reuteri* RC-14 similar to Kern et al. (2012).

Prebiotics, such as inulin, glycogen, or others, which support the metabolism of probiotics are sometimes added to probiotic tablets. However, within the field of gyneacolgy hitherto clinical studies to assess their superiority over probiotics are missing.

## 6.5 Summary and Conclusion

The normal and the abnormal vaginal microbiota is not yet fully understood. It is an ecosystem, which is influenced by genetic, ethnic, environmental and behavioral factors. More than 100 to 200 bacterial species, commensal, transient and endogenous, colonise the vagina and are influenced by the oral, rectal and penile microbiota. Cultural methods for the determination of normal or abnormal microbiota are insufficient and detect only a small, mostly aerobic, not representative and clinically unimportant spectrum. Lactobacilli mainly dominate the vaginal microbiota and are responsible, with other bacterial species, for the creation of a pH value between 3.8 and 4.5, which is considered as normal, at least in caucasian or asian women. Together with their antibacterial properties and immunological factors lactobacilli create a defense system against dysbiosis and infections within the vagina. This system is responsible for a healthy outer and inner genital tract, for a balanced restitution after intercourse and for a normal pregnancy and childbirth on time.

## References

Abdo CL, Safdar N (2009) The role of *Lactobacillus* probiotics in the treatment or prevention of urogenital infections – a systematic review. J Chemother 21:243–252

Adams MR (1999) Safety of industrial lactic acid bacteria. J Biotechnol 68:171–178

af Heurlin M (1914) Bakteriologische Untersuchungen des Genitalsekretes. Karger, Berlin

Amsel R, Totten PA, Spiegel CA, Chen KCS, Eschenbach DA (1983) Non specific vaginitis. Diagnostic criteria and microbial and epidemiologic associations. Am J Med 74:14–22

Antonio MA, Meyn LA, Murray PJ, Busse B, Hillier SL (2009) Vaginal colonization by probiotic *Lactobacillus crispatus* CTV-05 is decreased by sexual activity and endogenous Lactobacilli. J Infect Dis 199:1506–1513

Anukam K, Osazuwa E, Ahonkhai I, Ngwu M, Osemene G, Bruce AW, Reid G (2006) Augmentation of antimicrobial metronidazole therapy of bacterial vaginosis with oral probiotic *Lactobacillus rhamnosus* GR-1 and *Lactobacillus reuteri* RC-14: randomized, double-blind, placebo-controlled trial. Microbes Infect 8:1450–1454

Babula O, Lazdana G, Kroica J, Ledger WJ, Witkin SS (2003) Relation between recurrent vulvovaginal candidiasis, vaginal concentrations of mannose-binding lectin, and mannose-binding lectin gene polymorphism in Latvian women. Clin Infect Dis 37:733–737

Barrons R, Tassone D (2008) Use of *Lactobacillus* probiotics for bacterial genitourinary infections in women: a review. Clin Ther 30:453–468

Barton PT, Gerber S, Skupsky DW, Witkin SS (2003) Interleukin-1 receptor antagonist gene polymorphism, vaginal interleukin-1 receptor antagonist concentrations, and vaginal *Ureaplasma urealyticum* colonization in pregnant women. Infect Immun 71:271–274

Bohbot JM, Cardot JM (2012) Vaginal impact of the oral administration of total freeze-dried culture of LCR-35 in healthy women. Infect Dis Obstet Gynecol 2012: 503648

Cherpes TL, Hillier SL, Meyn LA, Busch JL, Krohn MA (2008) A delicate balance: risk factors for acquisition of bacterial vaginosis include sexual activity, absence of hydrogen peroxide-producing lactobacilli, black race, and positive herpes simplex virus type 2 serology. Sex Transm Dis 35:78–83

Cibley LJ, Cibley LJ (1991) Cytolytic vaginosis. Am J Obstet Gynecol 165:1245–1249

Coudeyras S, Marchandin H, Fajon C, Forestier C (2008a) Taxonomic and strain – specific identification oft he probiotic strain *Lactobacillus rhamnosus* 35 within the *Lactobacillus casei* group. Appl Environ Microbiol 74:2679–2689

Coudeyras S, Jugie G, Vermerie M, Forestier C (2008b) Adhesion of human probiotic *Lactobacillus rhamnosus* to cervical and vaginal cells and interaction with vaginosis-associated pathogens. Infect Dis Obstet Gynecol. doi:10.1155/2008/549640

Curtis AH (1913) A motile curved anaerobic bacillus in uterine discharges. J Infect Dis 13:165–169

de Vos WM, Engstrand L, Drago L, Reid G, Schauber J, Hay R, Mendling W, Schaller M, Spiller R, Gahan CG, Rowland I (2012) Human microbiota in health and disease. Self Care 3(S1):1–68

Demirezen S (2003) Cytolytic vaginosis: examination of 2947 vaginal smears. Cent Eur J Public Health 11:23–24

Döderlein A (1892) Das Scheidensekret und seine Bedeutung für das Puerperalfieber. Besold, Leipzig

Donders GG, Vereecken A, Bosmans E, Dekeersmaecker A, Salambier G, Spitz B (2002) Definition of a type of abnormal vaginal flora that is distinct from bacterial vaginosis: aerobic vaginitis. BJOG 109:34–43

Donders G, Bellen G, Rezeberga D (2011) Aerobic vaginitis in pregnancy. BJOG 118:1163–1170

Edwards L (2010) Dermatologic causes of vaginitis: a clinical review. Dermatol Clin 28:727–735

Ehrström S, Daroczy K, Rylander E, Samuelsson C, Johannesson U, anzén B, Pahlson C (2010) Lactic acid bacteria colonization and clinical outcome after probiotic supplementation in conventionally treated bacterial vaginosis and vulvovaginal candidosis. Microbes Infect 12:691–699

Eschenbach DA, Davick PR, Williams BC, Klebanoff SJ, Young-Smith K, Critchlow CM, Holmes KK (1989) Prevalence of hydrogen peroxid-producing *Lactobacillus* species in normal women and women with bacterial vaginosis. J Clin Microbiol 27:251–256

Falsen E, Pascual C, Sjoden B, Ohlen M, Collins MD (1999) Phenotypic and phylogenetic characterization of a novel *Lactobacillus* species from human sources: description of *Lactobacillus iners* sp. nov. Int J Syst Bacteriol 49:217–221

Fan A, Yue Y, Geng N, Zhang H, Wang Y, Xue F (2013) Aerobic vaginitis and mixed infections: comparison of clinical and laboratory findings. Arch Gynecol Obstet 287:329–335

Fettweis JM, Serrano MG, Girerd PH, Jefferson PH, Buck GA (2012) A new era oft he vaginal microbiome: advances using next-generation sequencing. Chem Biodivers 9:965–976

Fredricks DM, Fiedler TL, Marazzo JM (2005) Molecular identification of bacteria associated with bacterial vaginosis. N Engl J Med 353:1899–1910

Gajer P, Brotman RM, Bai G, Sakamoto J, Schütte UME, Zhong X, Koenig SSK, Li F, Ma Z, Zhou X, Abdo Z, Forney LJ, Ravel J (2012) Temporal dynamics of the human vaginal microbiota. Sci Transl Med 4:132ra52. doi:10.1126/scitranslatmed.3003605

Gardner HL, Dukes CD (1955) *Haemophilus vaginals* vaginitis – a newly defined specific infection previously classified "non-specific" vaginitis. Am J Obstet Gynecol 69:962–976

Genc MR, Onderdonk A (2011) Endogenous bacterial flora in pregnant women and the influence of maternal genetic variation. BJOG 118:154–163

Gray RH, Kigozi G, Serwadda D, Makumbi F, Nalugoda F, Watya S, Moulton L, Chen MZ, Sewankambo NK, Kiwanuka N, Sempijja V, Lutalo T, Kagayi J, Wabwire-Mangen F, Ridzon R, Bacon M, Wawer MJ (2009) The effects of male circumcision on female partner's genital tract symptoms and vaginal infections in a randomized trial in Rakai, Uganda. Am J Obstet Gynecol 200:42e1–42e7

Greenwood JR, Pickett MJ (1980) Transfer of *Haemophilus vaginalis* Gaedner and Dukes to a new genus, *Gardnerella*: *G. vaginalis* (Gardner and Dukes) comb. nov. Int J Syst Bacteriol 30:170–178

Guédou FA, van Damme L, Deese J, Crucitti T, Becker M, Mirembe F, Solomon S, Alary M (2013) Behavioural and medical predictors of bacterial vaginosis recurrence among female sex workers: longitudinal analysis from a randomized controlled trial. BMC Infect Dis 13:208. doi:10.1186/1471-2334-13-208

Hemmerling A, Harrison W, Schroeder A, Park J, Korn A, Shiboski S, Foster-Rosales A, Cohen CR (2010) Phase 2a study assessing colonization efficiency, safety, and acceptability of *Lactobacillus crispatus* CTV-05 in women with bacterial vaginosis. Sex Transm Dis 37:745–750

Hickley RJ, Zhou X, Pierson JD, Ravel J, Forney LJ (2012) Understanding vaginal microbiome complexity from an ecological perspective. Transl Res 160:267–282

Hillier SL, Meyn L, Macio I, Antonio M, Rabe L (2010) The back door reservoir for Lactobacilli and risk of BV (bacterial vaginosis) acquisition. Infect Dis Soc Obstet Gynecol. In: Proceedings of the 37th Annual Scientific Meeting, August 5, Santa Fe, New Mexico

Hilton E, Isenberg HD, Alperstein P, France K, Borenstein M (1992) Ingestion of yoghurt containing *Lactobacillus acidophilus* as prophylaxis for candidal vaginitis. Ann Intern Med 116:353–357

Homayouni A, Bastani P, Ziyadi S, Mohammad-Alizadeh-Charandabi S, Ghalibaf M, Mortazavian AM, Mehrabany EV (2014) Effects of probiotics on the recurrence of bacterial vaginosis: a review. J Low Genit Tract Dis 18; 79–86

Horowitz BJ, Mardh PA, Nagy E, Rank EL (1994) Vaginal lactobacillosis. Am J Obstet Gynecol 170:857–861

Hoyme UB, Hübner J (2010) Prevention of preterm birth is possible by vaginal pH screening, early diagnosis of bacterial vaginosis or abnormal vaginal flora and treatment. Gynecol Obstet Invest 70:286–290

Huang H, Song L, Zhao W (2013) Effects of probiotics for bacterial vaginosis in adult women: a meta-analysis of randomized clinical trials. Arch Gynecol Obstet 289: 1225–1234

Jahic M, Mulavdic M, Hadzimehmedovic A, Jahic E (2013) Association between aerobic vaginitis, bacterial vaginosis and squamous intraepithelial lesion of low grade. Med Arh 67:94–96

Jayaprakash TP, Schellenberg JJ, Hill JE (2012) Resolution and character-ization of distinct cpn60-based subgroups of gardnerella vaginalis in the vaginal microbiota. PLoS One 7(8):e43009. doi:10.1371/journal.pone.0043009

Jespers V, Menten J, Smet H, Poradosú S, Abdellati S, Verhelst R, Hardy L, Buvé A, Crucitti T (2012) Quantification of bacterial species of the vaginal microbiome in different groups of women, using nucleid acid amplification tests. BMC Microbiol 12:83

Jirovec O, Peter R, Malek J (1948) Neue Klassifikation der Vaginalbiocoenose auf sechs Grundbildern. Gynaecologia (Basel) 126:77

Kern AM, Bohbot JM, Cardot JM (2012) Traitement préventive de la candidose vulvovaginale récidivante par probiotique vaginal: résultats de l'étude observationelle Candiflore. Lett Gynécol 370:34–37

Köhler GA, Assefa S, Reid G (2012) Probiotic interference of *Lactobacillus rhamnosus* GR-*1* and *Lactobacillus reuteri* RC-14 with the opportunistic fungal pathogen *Candida albicans*. Infect Dis Obstet Gynecol 2012:636474. doi:10.1155/2012/636474

Krönig B (1895) Über die Natur der Scheidenkeime, speciell über das Vorkommen anaerober Streptokokken im scheidensekret Schwangerer. Centrbl Gynäkol 16:409–412

Lamont RF, Sobel JD, Akins RA, Hassan SS, Chaiworapongsa T, Kusanovic JP, Romero R (2011) The vaginal microbiome: new information about the genital tract flora using molecular based techniques. BJOG 118:533–549

Li J, McCormick J, Bocking A, Reid G (2012) Importance of vaginal microbes in reproductive health. Reprod Sci 19:235–242

Linhares IM, Giraldo PC, Baracat EC (2010) New findings about vaginal bacterial flora. Rev Assoc Med Bras 56: doi 10.1590/S0104-42302100000300026

Machado A, Jefferson KK, Cerca N (2013) Interactions between *Lactobacillus crispatus* and bacterial vaginosis (BV)-associated bacterial species in initial attachment and biofilm formation. Int J Mol Sci 14:12004–12012

Manhart MC, Fiedler TK, Fredricks DN, Marrazzo J (2012) Behavioral predictors of colonization with Lactobacillus crispatus or Lactobacillus jensenii after treatment for bacterial vaginosis: a cohort study. Infect Dis Obstet Gynecol 2012:706540. doi:10.1155/1012/706540. Epub 2012 May 30

Marrazzo JM, Fiedler TL, Srinivasan S, Thomas KK, Liu C, Ko D, Xie H, Saracino M, Fredricks DN (2012) Extravaginal reservoirs of vaginal bacteria as risk factors for incident bacterial vaginosis. J Infect Dis 205:1580–1588

Martin DH (2012) The microbiota of the vagina and its influence on women's health and disease. Am J Med Sci 343:2–9
Martin R, Langa S, Reviriego C, Jiminez E, Marin M, Xaus J, Fernandez L, Rodriguez J (2003) Human milk as a source of lactic acid bacteria for the infant gut. J Pediatr 143:754–758
Martinez RC, Seney SL, Summers KL, Nomizo A, de Martinis EC, Reid G (2009a) Effect of *Lactobacillus rhamnosus* GR-1 and *Lactobacillus reuteri* RC-14 on the ability of *Candida albicans* to infect cells and induce inflammation. Microbiol Immunol 53:487–495
Martinez RCR, Franceschini SA, Patta MC, Quintana SM, Candido RC, Ferrera JC, de Martinis ECP, Reid G (2009b) Improved treatment of vulvovaginal candidiasis with fluconazole plus probiotic *Lactobacillus rhamnosus* GR-1 and *Lactobacillus reuteri* RC-14. Lett Appl Microbiol 48:269–274
McMillan A, Dell M, Zellar MP, Cribby S, Martz S, Hong E, Abbas A, Dang T, Miller W, Reid G (2011) Disruption of urogenital biofilms by lactobacilli. Colloids Surf B Biointerfaces 86:58–64
Mendling W, Martius J, Hoyme UB (2013) S1 – guideline on bacterial vaginosis in gynecology and obstetrics. Geburtshilfe Frauenheilkd 73:1–4
Mirmonsef P, Gelbert D, Zariffard MR, Hamaker BR, Kaur A, Landay AL, Spear GT (2011) The effects of commensal bacteria on innate immune responses in the female genital tract. Am J Reprod Immunol 65:190–195
Mitchell C, Gottsch ML, Liu C, Fredricks DN, Nelson DB (2013) Association between vaginal bacteria and levels of vaginal defensins in pregnant women. Am J Obstet Gynecol 208:132.e1–132.e7
Nivoliez A, Camarez o, Paquet-Gachinat M, Bornes S (2012) Influence of manufacturing processes on in vitro properties of the probiotic strain *Lactobacillus rhamnosus* Lcr35. J Biotechnol 160:236–241
Nugent RP, Krohn MA, Hillier SL (1991) Reliability of diagnosing bacterial vaginosis is improved by a standardized method of gram stain interpretation. J Clin Pathol 29:297–300
Nürnberger L (1930) Die Erkrankungen der Scheide. In: Stoeckel (ed) Handbuch der Gynäkologie, revised 3rd edition of Handbuch der Gynäkologie by Veit J, vol 5. Bergmann, Munich
Petricevic L, Witt A (2008) The role of *Lactobacillus casei rhamnosus* Lcr35 in restoring the normal vaginal flora after antibiotic treatment of bacterial vaginosis. BJOG 115:1369–1374
Petricevic L, Unger FM, Viernstein Kiss H (2008) Randomized, double-blind, placebo-controlled study of oral lactobacilli to improve the vaginal flora of postmenopausal women. Eur J Obstet Gynecol Reprod Biol 141:54–57
Petricevic L, Domig KJ, Nierscher FJ, Krondorfer I, Janitschek C, Kneifel W, Kiss H (2012) Characterisation oft he oral, vaginal and rectal *Lactobacillus* flora in healthy pregnant and postmenopausal women. Eur J Obstet Gynecol Reprod Biol 160:93–99
Pirotta M, Gunn J, Chondros P, Grover S, O'Malley P, Hurley S, Garland S (2004) Effect of *lactobacillus* in preventing post-antibiotic vulvovaginal candidiasis: a randomised controlled trial. BMJ 329:458–51
Price LB, Liu CM, Johnson KE, Aziz M, Lau MK, Bowers J, Ravel J, Keim PS, Serwadda D, Wawer MJ, Gray RH (2010) The effects of circumcision on the penis microbiome. PLoS One 5:e8422
Ravel J, Gajer P, Abdo Z, Schneider GM, Koenig SSK, McCulle SL, Karlebach S, Gorle R, Russell J, Tacket CO, Brotman RM, Davis CC, Ault K, Peralta L, Forney LJ (2011) Vaginal microbiome of reproductive-age women. Proc Natl Acad Sci U S A 108(S1):4680–4687
Reid G (2001) Probiotic agents to protect the urogenital tract against infection. Am J Clin Nutr 73(Suppl):S437–S443
Reid G (2012) Probiotic and prebiotic applications for vaginal health. J AOAC Int 95:31–34
Rodriguez JM, Collins MD, Sjoden B, Falsen E (1999) Characterization of a novel *Atopobium* isolate from the human vagina: description of *Atopobium vaginae* sp. nov. Int J Syst Bacteriol 49:1573–1576
Sanchez D, Wagener J, Schaller M (2008) Impact of *Lactobacillus* species on localised *Candida albicans* infection and the mucosal innate immune response. Mycoses 51:434
Sanu O, Lamont RF (2011) Periodontal disease and bacterial vaginosis as genetic and environmental markers for the risk of spontaneous preterm labor and preterm birth. J Matern-Fetal Neonatal Med 24:1476–1485
Schaller M (2012) Lactobacilli in mucosal *Candida albicans* infections. In: de Vos WM, Engstrand L, Drago L, Reid G, Schauber J, Hay R, Mendling W, Schaller M, Spiller R, Gahan CG, Rowland I (eds) Human microbiota in health and disease. SelfCare 3(S1):S41–S45
Schröder R (1921) Zur Pathogenese und Klinik des vaginalen Fluors. Zbl Gynäkol 38:1350–1361
Senok AC, Verstraelen H, Temmerman M, Botta GA (2009) Probiotics for the treatment of bacterial vaginosis. Cochrane Database Syst Rev 7(4):CD006289
Shalev E, Battino S, Weiner E, Colodner R, Keness Y (1996) Ingestion of yogurt containing *Lactobacillus acidophilus* compared with pasteurized yogurt as prophylaxis for recurrent candidal vaginitis and bacterial vaginosis. Arch Fam Med 5:593–596
Sherman P, Ossa JC, Johnson-Henry K (2009) Unraveling mechanisms of action of probiotics. Nutr Clin Pract 24:10–14
Spiegel CA, Roberts M (1984) Mobiluncus gen. nov. Mobiluncus curtisii subsp. curtisii sp. nov., Mobiluncus curtisii subsp. holmesi subsp. nov., and Mobiluncus mulieris sp. nov., curved rods from the human vagina. Int Syst Bacteriol 34:177–184

Spurbeck RR, Arvidson CG (2011) Lactobacilli at the front line of defense against vaginally acquired infections. Future Microbiol 6:567–582

Strus M, Chmielarczyk A, Kochan P, Adamski P, Chelmicki Z, Chelmicki A, Palucha A, Heczko P (2012) Studies on the effect of probiotic *Lactobacillus* mixture given orally on vaginal and rectal colonization and on parameters of vaginal health in women with intermediate vaginal flora. Eur J Obstet Gynecol Reprod Biol 163:210–215

Swidsinski A, Mendling W, Loening-Baucke V, Ladhoff A, Swidsinski S, Hale LP, Lochs H (2005) Adherent biofilms in bacterial vaginosis. Obstet Gynecol 106:1013–1023

Swidsinski A, Mendling W, Loening-Baucke E, Swidsinki S, Dörfel Y, Scholze J, Lochs H, Verstraelen H (2008) An adherent *Gardnerella vaginalis* biofilm persists on the vaginal epithelium following standard therapy of bacterial vaginosis with oral metronidazole. Am J Obstet Gynecol 198:97e1–97e6

Swidsinski A, Dörffel Y, Loening-Baucke V, Mendling W, Schilling J, Patterson JL, Verstraelen H (2010) Dissimilarity in the occurrence of *Bifidobacteriaceae* in the vaginal and peranal microbiota in women with bacterial vaginosis. Anaerobe 16:478–482

Swidsinski A, Loening-Baucke V, Mendling W, Dörffel Y, Schilling J, Halwani Z, Jiang XF, Verstraelen H, Swidsinski S (2013) Infection through structured polymicrobial Gardnerella biofilms (StPM-GB). Histol Histopathol 29:567–597

Thies FL, König W, König B (2007) Rapid characterization of the normal and disturbed vaginal microbiota by application of 16S rRNA gene terminal RFLP fingerprinting. J Med Microbiol 56:755–761

Thomas S (1928) Döderlein's bacillus: *Lactobacillus acidophilus*. J Infect Dis 43:218–227

Vasquez A, Jakobsson T, Ahrné S, Forsum U, Maey olin G (2002) Vaginal *Lactobacillus* flora of healthy Swedish women. J Clin Microbiol 40:2746–2749

Verhelst R, Verstraelen H, Claeys G, Verschraegen G, Delanghe J, van Simaey L, de Ganck C, Temmermann M, Vaneechoutte M (2004) Cloning of 16S rRNA genes amplified from normal and disturbed vaginal microflora suggests a strong association between *Atopobium vaginae*, Gardnerella vaginalis and bacterial vaginosis. BMC Microbiol 4:16

Verhoeven V, renard N, Makar A, van Royen P, Bogers JP, Lardon F, Peeters M, Baay M (2013) Probiotics enhance the clearance of human papillomavirus-related cervical lesions: a prospective controlled pilot study. Eur J Cancer Prev 22:46–51

Verstaelen H, Verhelst R, Nuytinck L, Roelens K, de Meester E, de Vos D, van Thielen M, Rossau R, Delva W, de Bakker E, Vaneechoutte M, Temmermann M (2009) Gene polymorphisms of Toll-like and related recognition receptors in relation to the vaginal carriage of *G.vaginalis* and *A. vaginae*. J Reprod Immunol 79:163–173

Verstraelen H, Swidsinski A (2013) The biofilm in bacterial vaginosis: implications for epidemiology, diagnosis and treatment. Curr Opin Infect Dis 26:86–89

Verstraelen H, Verhelst R, Claeys G, de Bakker E, Temmermann M, Vaneechoutte M (2009) Longitudinal analysis of the vaginal microflora in pregnancy suggests that L. crispatus promotes stability of the normal vaginal microflora and that L. gasseri and/or L. iners are more conducive to the occurrence of abnormal vaginal microflora. BMC Microbiol 9:116. doi:10.1186/1471-2180-9-116

Weström L (Working Group Co-ordinator), Evaldson G, Holmes KK, van der Meijden W, Rylander E, Fredriksson B (1984) Taxonomy of bacterial vaginosis; bacterial vaginosis – a definition. Symposium on bacterial vaginosis, Stockholm, Jan. 1984. In: Mardh PA, Taylor-Robinson D (eds) Bacterial vaginosis. Almquist & Wiksell, Stockholm, pp 259–260

Ya W, Reifer C, Miller LE (2010) Efficacy of vaginal probiotic capsules for recurrent bacterial vaginosis: a double-blind, randomized, placebo-controlled study. Am J Obstet Gynecol 203:120.e1–120.e6

Zhou X, Bent SJ, Schneider MG, Davis CC, Islam MR, Forney LJ (2004) Characterization of vaginal microbial communities in adult healthy women using cultivation-independent methods. Microbiology 150:2565–2573

Zhou X, Hansmann MA, Davis CC, Siuzuki H, Brown CJ, Schütte U, Pierson JD, Forney LJ (2010) The vaginal bacterial communities of Japanese women resemble those of women in other racial groups. FEMS Immunol Med Microbiol 58:169–181

# The Human Gut Microbiota 7

Hermie J.M. Harmsen and Marcus. C. de Goffau

**Abstract**

The microbiota in our gut performs many different essential functions that help us to stay healthy. These functions include vitamin production, regulation of lipid metabolism and short chain fatty acid production as fuel for epithelial cells and regulation of gene expression. There is a very numerous and diverse microbial community present in the gut, especially in the colon, with reported numbers of species that vary between 400 and 1500, for some those we even do not yet have culture representatives.

A healthy gut microbiota is important for maintaining a healthy host. An aberrant microbiota can cause diseases of different nature and at different ages ranging from allergies at early age to IBD in young adults. This shows that our gut microbiota needs to be treated well to stay healthy. In this chapter we describe what we consider a healthy microbiota and discuss what the role of the microbiota is in various diseases. Research into these described dysbiosis conditions could lead to new strategies for treatment and/or management of our microbiota to improve health.

**Keywords**

Microbiota • Development • Obesity • Aberrant microbiota • Diabetes

## 7.1 Introduction

The microbiota in our gut normally helps us to stay healthy (Hooper and Gordon 2001; Sekirov et al. 2010). It performs many different functions that are essential for us, such as vitamin production, detoxification of toxins, regulation of cholesterol metabolism, bile deconjugation, providing colonization resistance to pathogens, SCFA production as fuel for epithelial cells and regulation of gene expression (Walter and Ley 2011). However, the main function of the microbe itself is to survive and thrive. The microbe is there because, maybe by coincidence, it has found a niche to grow in. Whatever these microbes do, they need to be able to colonize our gut, to grow on the nutrition provided and to reproduce in a rapid way, because in, let's say

H.J.M. Harmsen (✉)
Department of Medical Microbiology, University of Groningen, University Medical Center Groningen, 30001, 9700 Groningen, The Netherlands
e-mail: H.J.M.Harmsen@med.umcg.nl

M.C. de Goffau
Wellcome Trust Sanger Institute, Hixton, CB10 1SA Cambridge, UK



A. Schwiertz (ed.), *Microbiota of the Human Body*, Advances in Experimental Medicine and Biology 902, DOI 10.1007/978-3-319-31248-4_7

24–36 h (Heaton et al. 1992), the bowel is emptied and the cycle starts again. The microbiota in our gut performs many different essential functions that help us to stay healthy. These functions include vitamin production, regulation of lipid metabolism and short chain fatty acid production as fuel for epithelial cells and regulation of gene expression. There is a very numerous and diverse microbial community excisting in the gut, especially in the colon, with reported numbers of species that vary between 400 and 1500, for some those we even do not yet have culture representatives.A healthy gut microbiota is important for maintaining a healthy host. An aberrant microbiota can cause diseases of different nature and at different ages ranging from allergies at early age to IBD in young adults. This shows that our gut microbiota needs to be treated well to stay healthy. In this chapter we describe what we consider a healthy microbiota and discuss what the role of the microbiota is in various diseases. Research into these described dysbiosis conditions could lead to new strategies for treatment and/or management of our microbiota to improve health.

The presence of a very numerous and diverse microbial community excisting in the gut, especially in the colon, is thus quite amazing. The numbers of different species that are mentioned in the current studies vary between 400 and 1,500 species (Qin et al. 2010; Turnbaugh et al. 2010; Gill et al. 2006). These species are mostly expressed as operational taxonomic units (OTU's), these are unique sequence types that should represent a species but do not always have a specific name yet, since there is no culture representative yet or the taxonomy is lacking behind (Rajilić-Stojanović and Vos 2014). The most dominant species present in healthy individuals are illustrated in Fig. 7.1 with an indication of the substrate utilization and the relative abundance in human feces as detected in different next generation sequencing studies (Flint et al. 2014; Qin et al. 2010; David et al. 2013).

The bacteria taxonomy can be quite confusing some way. As an example we will explain the taxonomy of the well-known gut bacterium *Escherichia coli*: the double name according Linnaeus shows first the genus name *Escherichia*, followed the species name *coli*. It belongs to the family of *Enterobacteriaceae*, part of the order of *Enterobacteriales*, which is part of the class of Gammaproteobacteria belonging to the phylum of Proteobacteria of the kingdom Bacteria (Whitman and Parte 2009; Moore et al. 1987).

There are four numerically important bacterial phyla present in the adult human gut: Bacteriodetes (Gram-negative anaerobes) and Firmicutes (Gram positives) are most dominant, followed by Actinobacteria and Proteobacteria. Major Bacteriodetes genera in the gut are *Bacteroides* and *Prevotella*, major Firmicutes genera are *Clostridium*, *Blautia*, *Faecalibacterium*, *Eubacterium*, *Roseburium*, *Ruminococcus*, *Streptococcus and Lactobacillus*. Actinobacteria are represented by the genera *Bifidobacteria*, *Atopobium* and *Collinsella*, the Proteobacteria by *Enterobacteriaceae* like the genus *Escherichia* (Tap et al. 2009; Walker et al. 2010). An additional phylum the Verrucomicrobia is represented by one species, *Akkermansia muciniphila* (Derrien et al. 2004). This is a very simplistic overview, in reality the families present in the gut are subdivided by a long list of well-known genera, but also new genera, such as *Christensenella* (Morotomi et al. 2012) and reclassified genera, such as *Erysipelatoclostridium ramosum*, a reclassification of *Clostridium ramosum* (Yutin and Galperin 2013). Despite the current advances in detection and identification techniques, it will take some time until a stable taxonomic list is created which contains a proper naming for all the species present in the gut. Furthermore, next to bacteria there are also members of the kingdom Archaea present, predominantly *Methanobrevibacter* species that produce methane in the gut (Samuel et al. 2007), and Eukarya, such as the yeast *Candida*, microbial parasites, such as *Entamoeba* (Parfrey et al. 2011) and macro parasites, for example helminths (Weinstock and Elliott 2009) . Finally, also viruses and bacteriophages play an significant role in the ecology and maintenance of a healthy balance in the gut (Kernbauer et al. 2014).

The driving force for the maintenance of the complex microbiota in our gut should be sought

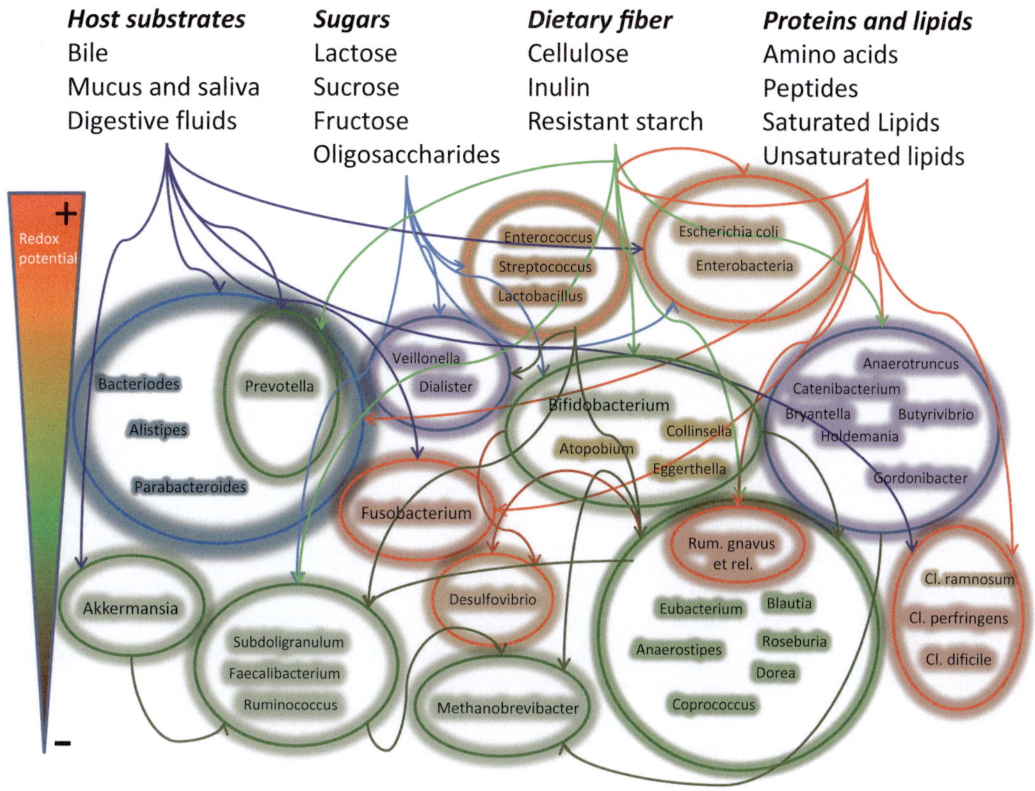


**Fig. 7.1** Diagram with dominant bacterial genera and species in relation to the preferred redox potential and the substrate utilization in the human gut. Colors correspond to the associations with health (*green*), disease (*red*) or a dual (*blue*) or unclear role (*purple*). Size of the circles reflect the average abundance in the gut. *Rum* is Ruminococcus, *Cl* is clostridium

in the complexity of substrate availability and, perhaps even more importantly, the (mutualistic) interactions between the many different species (Flint et al. 2012a; Ley et al. 2008; Arumugam et al. 2011). Each species in the gut fills up a specific niche, which could be best described by the substrates these bacteria utilize. If no external nutrition would enter the gut, the only substrate bacteria would have are host excreta, such as saliva, mucus, gastric and pancreatic juices and bile. Most of the excreta create harsh conditions, like a low pH, contain proteolytic enzymes, or anti-microbial substances. Mucins from saliva and mucus on the intestinal epithelium, will most likely be the most dominant substrates for bacteria to utilize (Arumugam et al. 2011). The phylum Bacteroidetes including *Prevotella* and *Bacteriodes* species are versatile bacteria that have the capabilities to utilize mucins with their large spectrum of carbohydrolases (Ouwerkerk et al. 2013; Benjdia et al. 2011). *Akkermansia muciniphila* is a bacterium specialized in degrading mucins and will be part of this niche (Derrien et al. 2004). Cross feeding and syntrophic interactions will enable other microbes to utilize the carbohydrates liberated from these mucins, and the various proteins and amino acids from the excreta. For instance, *Faecalibacterium prausnitzii* is able to utilize the N-acetyl-D-glucosamine liberated from mucins (Wrzosek et al. 2013; Ouwerkerk et al. 2013; Lopez-Siles et al. 2012). Enterobacteria will also be able to grow on left over peptides and carbohydrates, and will utilize the low amount of oxygen and other electron acceptors penetrating from the epithelial lining (Walter and Ley 2011). The growth on excreta alone will already sustain a diverse population in gut. This diversity will multiply when nutrition comes in that escaped host's digestion of the diet. Especially, soluble and insoluble dietary fiber will enable the growth of many typical gut anaerobes that ferment these into short chain fatty acids (Flint et al. 2012b). These will be *Prevotalla* and *Bacterodes* species, ruminococci, bifidobacteria, lactic acid bacteria, eubacteria and clostridia as well as fungi like saccharomyces and candida (Martens et al. 2009; Ze et al. 2012; Flint et al. 2008).

Figure 7.1 shows presumptive utilization of the main nutrients available for the gut microbiota. From this we can see that there is a lot of redundancy in the metabolic capacities of the microbes. This may explain the variation and diversity of microbiota in different individuals.

In 2011 the concept of enterotypes was introduced, which proposed that the gut microbiota of all human could be divided in at least three enterotypes (Arumugam et al. 2011). These enterotypes are Bacteroides type, the Prevotella type and the Ruminococcus type according to the central microbe in an association network. The authors state that "the enterotypes are in fact driven by groups of species that together contribute to the preferred community compositions". This is based on the core consortium of microorganisms that breaks down the main complex substrates and especially mucin degradation seems to be an important determinant (Arumugam et al. 2011). This study was followed by the study of Wu which showed that enterotypes were strongly associated with long-term diets, particularly protein and animal fat (Bacteroides) versus carbohydrates (Prevotella) (Wu et al. 2011) which could be influenced by diet (David et al. 2013).

We believe that a well-functioning gut microbiota helps to ensure that the gut remains healthy and is therefore important in maintaining the health of the host. When there is in a unbalanced state (dysbiosis of microbiota), the functioning of the gut microbiota will be impaired and this may lead to disease. Current developments in gut microbiota analysis techniques have opened up the possibility to show clear associations in between particular gut microbial compositions (or the lack thereof) and the development of various diseases. In this chapter we want to show how diseases can be link with and dysbiosis and malfunctioning of the gut microbiota. The main diseases that are clearly linked to an aberrant gut microbiota are inflammatory bowel diseases (IBD) like Crohn's disease and ulcerative colitis. Another category of diseases are the autoimmune diseases like atopy, eczema, asthma, celiac disease and type 1 diabetes. A last category that will be discussed here are related to an increasing welfare and age: obesity, metabolic disease and type 2 diabetes.

## 7.2 The Normal Microbiota

The normal gut microbiota is any microbiota that is present in the gut of a healthy individual. Various studies have shown that there is a large inter-individual variation, no gut microbiota is the same. There is a relation with genetics as identical twins have a more similar microbiota than siblings, which in turn share more similarities with one another than not related people (Turnbaugh et al. 2010; Zoetendal et al. 2001; Tims et al. 2012). There is some relation with cultural habits and diets in different regions of the World. However, a lot of variation is not explained and this will have to do with co-incidental colonization and stabilization of an ecosystem in an existing niche that is formed by genetics, diet, age and habits. Despite this variation there is a common functionality of all these microbiota's, being the breakdown of the available nutrients.

This commonality will cause a division of the microbiota's in different types, defined as enterotypes before (Arumugam et al. 2011). In this chapter we would like to share a few thoughts on these enterotypes based on what we see in various studies described in literature combined with our experiences. We see four types of microbiota:

The first type is the common type in the North-West European regions. It seems to be related with an omnivore type diet, based on grains, fruits and vegetables, and meat. This is a type dominated by bacteria of *Clostridium* clusters IV and XIVa, clusters containing many butyrate- and other SCFA-producing bacteria. Especially the number of *Faecalibacterium* are high in this type. Furthermore, there is a fair amount of bifidobacteria and mucus degrading *Akkermansia*. *Bacteroides* are present in reasonable numbers but not very dominant and there is a low level of *E. coli* and relatives (de Goffau et al. 2014; Harmsen et al. 2002).

The second type is more related to a fibre rich (vegetarian) diet with little or no animal products. Found in a part of the western population, but also in for instance, in rural Africa (de Vrieze 2014; De Filippo et al. 2010). This type is dominated by *Prevotella* and *Dialister-Veillonella* bacteria. There are also members of *Clostridium* clusters IV, especially ruminococci, and a little XIVa. It characterizes itself by the very low amount of *Bacteroides* and low *E. coli*.

The third type is a type found especially in eaters of an western type diet high in animal protein and fat. This type is dominated by *Bacteroides*, and high levels of *Clostridium* clusters IV *Faecalibacterium* and of *Clostridium* clusters XIVa *Ruminococcus gnavus* (David et al. 2013). This type has virtually no *E. coli*.

The last type we find is a type that seems to be in dysbiosis, since it is often found in relation to inflammatory diseases and diarrhea. It is characterized by a high amount of enterobacteria, such as *E. coli*. Furthermore, *R. gnavus* is relatively high and there are *Bacteroides* present. There is a low amount of other *Clostridium* clusters XIVa species, and a low amount of *Clostridium* clusters IV (Harmsen et al. 2012; Willing et al. 2009; Morgan et al. 2012). This very rough description seems to be useful if we relate it to function and disease. The first two seem to be healthy microbiota's, the third *Bactoroides* type seems a risk microbiota and the last type a clear microbiota "out of order". Although the latter may be a temporarily state, for instance caused by enteroviral infection, it does seem to be a non-stable type vulnerable to chronic dysbiosis.

## 7.3 Gut Microbiota Development: The Young Microbiota

The gut of an unborn child is in principle sterile in the womb and microbiota development starts as soon as a child passes the birth channel. The colonization starts with a diverse microbiota of the mother, where *Bacteroides/Prevotella*, lactobacilli, streptococci, staphylococci, and *Enterobacteriaceae* species grow out quickly to dominant numbers (Penders et al. 2006; Isolauri 2012; Dominguez-Bello et al. 2010; Harmsen et al. 2000). When nutrition starts the microbiota will change towards those species that are selected by the nutrition. In most cases the nutrition will be breastfeeding that contains bacteria as well to colonized the baby (Harmsen et al. 2000; Martin et al. 2009). Components of breast milk are strongly bifidogenic. This is not primarily lactose but also specific oligosaccharides for which bifidobacteria have specific enzymes (Sela et al. 2008; Zivkovic et al. 2011). This normally results in a dominance of bifidobacteria within 1 week after birth in some cases this can be almost up to 100 % bifidobacteria in the gut microbiota. As the child gets older the bifido-dominated microbiota gets more diversified. It is not clear what is driving this but gastrointestinal infections and antibiotic use will play a role in this. The species of bifidobacteria that colonize the gut may vary a lot between individuals, although *B. longum* seems to be most common combined with colonization of either *B. bifidum*, *B. breve*, and at a later time point *B. pseudocatenulatum* or *B. adolescentis* (Bergstrom et al. 2014). Between 3 and 6 months introduction of foods other than milk usually starts. Often the solid foods are fruits and roots, such as carrot, followed by peas and beans etc. This means a nutrition driven diversification of the microbiota. With the introduction of these solid foods also dietary fibre degrading bacteria will colonize, such as ruminococci, *Bacteroides*, *Prevotella* and *Clostridium* clusters IV and XIVa (Favier et al. 2002). Also other firmicutes species and different enterobacteria will become introduced. This weaning period is a critical phase in the colonization of the children as it coincides with a period in time of immune system training (Grönlund et al. 2007).

At the age of 1 year the microbiota appears to have the major functions that an adult microbiota should have, however the microbiota seems to increasingly diversify until the age of 3 years where it starts to stabilize (Yatsunenko et al. 2012). This stabilization leads to a phase where the microbiota does slowly develops further. The microbiota of young children and adolescents is still different from that of adults and the microbiota continues to develop and change until old age, where there are again significant differences with the microbiota of e.g. 30 year old adults (Claesson et al. 2011). Many factors play a role in the early development of the microbiota. Breastfeeding vs. formula feeding and Caesarian section vs. natural delivery are logical important determinants of the gut microbial development which seem to have a long term effect (Lozupone et al. 2013; Adlerberth and Wold 2009; Van Nimwegen et al. 2011). Less obvious factors such as maternal factors expressed in breast milk, host factors like genetics and innate immunity are however also important (Martín et al. 2007; Zivkovic et al. 2011). Environmental factors and choice and timing of nutrition subsequently continue to steer the development of the microbiota. There are of course a lot of environmental factors that can influence the microbiota development. Environment can be determined by conditions that favor encounters with different micro-organisms, such as hygiene conditions, number of siblings, daycare centers and school, animals in the house, rural or urban lifestyle (Azad et al. 2013). These condition will determine the composition of the gut microbiota, just as later in life meat, sugar and fat consumption, smoking and alcohol consumption (David et al. 2013). All these factors will be important in relation to diseases where the gut microbiota is believed to be involved (Lozupone et al. 2013). For example, atopy and eczema, which are correlated with a dysbiosis of the microbiota, are also negatively correlated with the number of siblings (Penders et al. 2007).

These relations can be explained by the hygiene hypothesis that state that we have to encounter sufficient microbes for the programming of our immune system. Extra siblings bring extra microbes with them that helps to diversify the microbiota. The hygiene hypothesis is currently tested in several studies, to find or exclude a relation with autoimmune diseases that are currently on the rise like atopy, asthma and type 1 diabetes (Vaarala et al. 2008; Dunne et al. 2014).

## 7.4 The Role of Gut Microbiota in Disease

As stated before, gut microbiota is involved in health and when it is in a state of dysbiosis, the gut microbiota may lead to disease, although a causal relationship is often hard to show. Current developments in gut microbiota analysis revealed the involvement of microbiota in various diseases. The main diseases that are clearly linked to an aberrant gut microbiota are inflammatory bowel diseases (IBD) like Crohn's disease and ulcerative colitis. Another category of diseases are the auto-immune diseases like atopy eczema, asthma, celiac disease and type 1 diabetes. A last category that will be discussed here are related to an increasing welfare and age: obesity and type 2 diabetes.

### 7.4.1 IBD

Inflammatory bowel diseases were already associated with an aberrant gut microbiota for a long time ((Xavier and Podolsky 2007). Bacteria such as mycobacterium, adherent invasive *E. coli* (AIEC) and viruses have been associated with IBD (Fiocchi 1998). Today it seems apparent that IBD is not caused by a single micro-organism but that it is an delicate interplay between genetics, immunology, environmental factors and gut microbiota. Principal component analysis of data from metagenomic sequencing clearly showed that Crohn's disease patients had a microbiota that was different from healthy controls and UC patients, who's microbiome seems only moderately different from healthy patients (Qin et al. 2010). It shows first of all that CD and UC are different diseases, of which the microbiota of CD patients is clearly more affected than that of UC patients. What it doesn't show is what causes CD but merely what the effect of inflammation is on the microbiota and vice versa. Patients in remission still show a high number of enterobacteria and a reduced number of faecalibacteria and other butyrate producing species (Sokol et al. 2009; Miquel et al. 2013). An inverse relationship between the first two groups is seen in many of the studies on IBD (Willing et al. 2009; Harmsen et al. 2012). *Bacteroides* numbers are usually also higher in CD patients, although this may have to do with a lowering of other bacteria like *Clostridium* groups IV and XIVa. It seems that due to the inflammation, the gut lumen is less reduced so that facultative anaerobes and oxygen tolerant bacteria get more room to grow, hence the increase in enterobacteria and along with that, enterococci. On the other hand, strict anaerobes like faecalibacteria and other *Clostridium* group IV and XIVa will see their numbers diminish in such conditions. The mechanism behind this is the production of reactive oxygen species (ROS) and nitric oxide (NO·) during intestinal inflammation (Winter et al. 2013). For instance, NO-production by inducible nitric oxide synthase (iNOS) leads to the formation of nitrite formation in the gut. This will be used for anaerobic respiration by locally present facultative *Enterobacteriaceae* that in this way have a competitive advantage over the strict anaerobes that are troubled by the oxidative stress anyway. This effect also explains the loss of colonization resistance after antibiotic treatment. Experiments with streptomycin showed that antibiotic increased the epithelial inflammation and thus stimulated growth of enterobacteria by nitrate respiration rather than changing the microbial populations themselves (Spees et al. 2013).

A changed microbiota in CD patients would be a result of the chronic inflammation and not necessarily the cause. The cause may be purely coincidental, a viral infection, or infection with pathogenic *Enterobacteriaceae* or even a trauma such as disturbance of the intestinal integrity (Xavier and Podolsky 2007; Strober 2011; Liverani et al. 2014). This results in inflammation of the gut epithelium and in genetically-predisposed individuals this would lead to chronic inflammation, maybe because they may be impaired in autophagy or their anti-inflammatory signaling (Sadaghian Sadabad et al. 2014). Periods of inflammation could be followed by a dysbiosis of the gut microbiota which is unable to restore its balance. In that case, the microbiota does not produce the right SCFA and anti-inflammatory signals for the epithelial lining to regenerate itself and hence becomes susceptible to new inflammatory triggers (Sokol et al. 2008). This vicious cycle is considered in the concept of a leaky gut (Fasano 2012). A possible cure to break this cycle could on one hand be immunological, such as to steer the inflammatory signaling by improving the autophagy or regulate (anti)-inflammatory cytokines. Or on the other hand, try to improve the dysbiosis so that a balanced microbiota produces the right SCFA and anti-inflammatory signals, for a well-functioning gut barrier with a low grade of inflammation, properly responding to normal triggers. The focus is now on the use of anaerobic butyrate producers, such as faecalibacteria, as probiotics to restore the dysbiosis (Miquel et al. 2013). The role of the microbiota in UC patients may be more difficult to understand. With UC the gut is mildly inflamed but over a larger area compared to CD the gut microbiota is usually less affected. There seems to be a crucial role for the interplay between the adaptive immune system (neutrophyles) and the microbiota (Fries and Comunale 2011; Chen et al. 2014). Both IBD diseases affect usually young adults, these patients often had a normal gut function in their childhood. Whether their gut microbiota was effected before the disease set in, remains to be seen.

## 7.5 Abberant Gut Microbiota Development and Type 1 Diabetes

Auto-immune diseases correlate, unlike IBD, with an aberrant microbiota development. Various studies have been performed in humans and in rodents in relation to the development of type 1 diabetes (T1D) and its association with the gut microbiota. Many of the associations found between the gut microbiota and T1D overlap between the different studies and they importantly do not contradict one another. A coherent picture in which the gut inflammatory state is reversely coupled to the abundance of butyrate-producing bacteria, emerges consistently out of all these studies. Sufficient butyrate production keeps the gut healthy as it is the main energy source for colonic epithelial cells (Hague et al. 1996), regulates the assembly of tight junctions (reduces permeability) (Peng et al. 2009) and it importantly reduces intestinal inflammation (Hamer et al. 2009). Several bacterial genera from the *Clostridium* clusters IV and XIVa, such as *Faecalibacterium* and *Roseburia*, are well known butyrate producers and are usually under-represented in the microbiota of people that either have T1D or that are developing it (de Goffau et al. 2013). On the other hand, diabetogenic bacterial groups are also identified and interestingly, the *Bacteroides* genus, is found to be associated with T1D without exception in basically every study (de Goffau et al. 2014; Murri et al. 2013; Brugman et al. 2006; Giongo et al. 2010; Brown et al. 2011). Several explanations exist for this particular association with *Bacteroides*. The first reason is that *Bacteroides* is not a butyrate producer as it producers propionate and acetate; an overabundance of *Bacteroides* will leave less room for actual butyrate producers in the gut (competition for resources). Secondly, *Bacteroides*, like *Escherichia coli*, are Gram-negative bacteria of which particular cell wall components, namely the lipopolysaccharides (LPS), can induce inflammation by acting as endotoxins (inducing an inflammatory response via the immune system). A third possible link with T1D, is that *Bacteroides* (and streptococci) produces glutamate decarboxylase (GAD), which might be a trigger for GAD autoimmunity (GADA) via molecular mimicry. Glutamate decarboxylase is also produced in humans (GAD65) but only in the insulin producing beta cells of the pancreas and in neuron cells. When the human body raises autoantibodies against bacterial GAD it might accidentally subsequently attack its own GAD producing cells, hence resulting in T1D as the insulin producing cells are destroyed. Interestingly, Alzheimer's disease is also considered to be a type of diabetes (Type 3), or a result of it (Vignini et al. 2013), and this might in part also be due to the destruction of GAD producing neurons. *Bacteroides* can thus accommodate its translocation past the gut barrier where the immune system subsequently comes into contact with it by indirectly hampering the production of butyrate by other bacteria via competition and directly by inducing inflammation via LPS.

Rather large differences exist between countries in regards to the prevalence of T1D. Part of this can be explained genetically but a large part is likely due to lifestyle and nutrition patterns. The effects of diet on the prevalence of *Bacteroides* have especially been well documented as a western diet high in protein and animal fat consumption (think hamburgers and fast foods) leads to a high prevalence of *Bacteroides* (David et al. 2013). People in developing countries eating traditional foods on the other hand usually have a microbiota which is dominated by *Prevotella* instead (De Filippo et al. 2010) as *Prevotella* are mainly associated with a carbohydrate rich diet (roots, tubers, corn, sorghum, etc.). In western countries a healthy gut microbiota composition would be considered to either consist out of a composition which is largely dominated by bacteria from *Clostridium* clusters IV and XIVa, bifidobacteria and just small amounts of Bacteroides or a composition which is more centered around *Prevotella* (*Prevotella* enterotype), in which *Clostridium* cluster IV is still always present in high abundance but where *Bacteroides* is largely absent. Both these healthy types of microbiota composition are likely established by a diet rich in dietary fiber and a moderate consumption of animal products and limited amounts of (processed) sugars. An interesting example of the above is a study by Mejía-León et al. done in Mexico, where the incidence of T1D is higher in people living close to the border with the USA. Patients who just developed T1D in this study had higher *Bacteroides* numbers while the healthy controls had a gut microbiota which was dominated by *Prevotella. Bacteroides* is far more dominant in the USA than in Mexico and T1D rates are indeed much higher there (Mejía-León et al. 2014). Similar kinds of patterns can be discerned when comparing Finland with Estonia (De Goffau, EU project Diabimmune, unpublished).

## 7.6 Obesity and Metabolic Diseases

There is currently a lot of interest in research for obesity, a metabolic disease that became epidemic in the Western world (Ley et al. 2005). One of the first success stories of next-generation sequencing is the work of the lab of Gordon and co-workers on the role the microbiota in this disease. Their research suggested that next to a chronic energy surplus also other factors such as the indigenous gut microbiota could play a role (Backhed et al. 2004). Even more, that the energy overshoot could be caused by an overproduction of SCFA in the colon and that there is a low grade of inflammation that often symptomizes obesity and related metabolic diseases (Fleissner et al. 2010; Backhed et al. 2007).

Obesity can lead to serious comorbidities, such as type 2 diabetes and liver cirrhosis caused by alcoholic or nonalcoholic liver diseases. Recent studies show a dysbiosis in both type 2 diabetes and liver cirrhosis, although the nature and intensity of the dysbiosis is different. In type 2 diabetes there is only a mild dysbiosis, and fecal samples of the patients seem to be dominated by firmicutes other than the butyrate-producing *Roseburia* and *F. prausnitzii*, and *E. coli* (Qin et al. 2012; Khan et al. 2014). In liver cirrhosis the samples were dominated by *Bacteroides* and buccal microbiota (Qin et al. 2014). In this case the buccal microbiota may be able to colonize the ileal and colonic gut since there is a compromised bile production and less bile-resistant bacteria can survive.

Recent studies showed that lipopolysaccharides (LPS) cause the low grade inflammation in obesity, a metabolic endotoxemia that dysregulates the inflammatory tone and triggers body weight gain and diabetes (Cani et al. 2007). The scientists concluded that the system of CD14 immune cells reacting on LPS set the tone of insulin sensitivity and the onset of diabetes and obesity. At this moment it is clear that obesity is somehow associated with altered gut microbiota, low-grade inflammation and a cluster of metabolic disorders. By inducible intestinal epithelial cell-specific deletion of MyD88 in mice, Cani and his colleagues showed recently that this signal-transduction protein partially protects against diet-induced obesity, diabetes and inflammation (Everard et al. 2014). These mice spend more energy and had improved glucose homeostasis, reduced liver fattening, fat mass and inflammation. This partial protection could be transferred to germ-free mice by gut microbiota transplantation. Remarkably, the MyD88 deletion increased anti-inflammatory endocannabinoids, restored antimicrobial peptides production and increases intestinal regulatory T cells during diet-induced obesity. These experiments show that the intestinal epithelial MyD88 acts like a sensor switching the host metabolism according to the nutritional status. Obesity and related disorders may therefore be again a result of an unbalanced relationship between microbiota and epithelial innate immunity, regulating metabolic functions normally needed for optimal energy harvest for nutrient limited diet.

In metabolic diseases the microbiota is out of balance, maybe due to diet or all kinds of different co-morbidities, maybe due to excessive behavior, such as over eating, alcohol abuse or sleep deprivation (Cronise et al. 2014). Abnormal production of SCFA and other metabolites as well as an altered lipid metabolism seem to play a major role in the etiology of these diseases. Propionate over production by *Bacteroides* species and over production of acetate and butyrate by different firmicutes may lead to an energy surplus (Holmes et al. 2012; Turnbaugh and Gordon 2009; Schwiertz et al. 2010). Different microbial metabolites influence the satiety of the host and regulate the adipose tissue formation (Roelofsen et al. 2010). However, the effect of the gut microbiota may be secondary. Improving the obesity by special protein rich diets may change the microbiota, but not always for the better (Walker et al. 2010), since they lead to a more *Bacteroides* type of microbiota (David et al. 2013). A more sustainable diet with low calories and high fibre intake will lead to a more balanced microbiota, that can further improve the metabolic disease and will have a positive effect on the insulin resistance (Walker et al. 2010).

Fecal transplantation as a way to improve the obesity and type 2 diabetes has led to improved insulin resistance, but not immediately improves the BMI of the patients (Vrieze et al. 2012). Controlled diets may improve both, since this will automatically improve the gut microbiota as well. Furthermore, it seems unlikely that obesity is caused by microbes, but merely shows the effect of diet on the microbiota. New metabolomics and metagenomic studies will give a more clearer picture on the relationship between obesity and gut microbiota.

## 7.7 Concluding Remarks

A healthy gut microbiota is important for maintaining a healthy host. An aberrant microbiota seems to slowly cause diseases of different nature and at different ages. From allergies at early age to IBD in young adults, from autisms to colorectal cancer, a large variety of effects are reported. This results in an obvious need for gut microbiota management. Fortunately, it automatically goes alright in most of the people that have a healthy, regular lifestyle. But in case of dysbiosis and disease, modulation of the microbiota is wanted by means of pro- and prebiotics or in extreme cases by fecal transplantation. There are still some essential questions to be answered on the gut microbiota. We still do not really know what is a healthy microbiota, nor do we know what the influence of our genetic makeup is on the microbiota. What is a healthy diet? Are we not apes that should eat nuts, roots and fruits and occasionally a piece of meat? How would our gut microbiota look like if we would eat just that, would it look like ape microbiota and would that be healthy (Ellis et al. 2013)?

There are some clear generalities, such as: for a well-developed gut and immune system a diverse microbiota is needed, diversity in species as well as in gene richness. The production of SCFA in the right relative proportions is needed. However, there is a special role for butyrate, this may be very important for a healthy gut. A low butyrate production may lead to a poor nourished gut epithelium that opens its tight junctions and may become leaky, with different diseases as consequence. In this healthy microbiota there is therefore a role for butyrate producing bacteria. *F. prausnitzii* is a species which seems to have co-evolved with humans, since it is present in most humans at high numbers. Next to this species there is a role for *Roseburia*, *Clostridium* group XIVa bacteria that, like faecalibacteria use the butyryl-CoA pathway to produce butyrate from glucose and acetate. Also other types of butyrate producers seem to be important in this respect, such as *Coprococcus*, *Anaerostipes*, and *Eubacterium halii*, bacteria that produce butyrate from acetate and lactate.

Supplementing the microbiota with these micro-organisms may not be enough since these bacteria require low redox potentials to grow and imbedding in a complex trophic network. This may not always be provided by the host and the dysbiosed gut microbiota.

For most diseases where a relation with a dysbiosed microbiota is demonstrated, still the chicken and egg question has not been answered: a causal relationship has not been shown. What is the trigger of the deteriorating situation. Viruses may certainly have a role as trigger, just like yeast and parasites, which are not discussed in this chapter. The host may have undergone immunological changes that makes it susceptible to triggers from microbiota to start aberrant immune reactions.

Much needs to be investigated and this makes us realize that we are just in the beginning of understanding the microbial world within us an its relation the health and disease. Fortunately, new system biology and meta-omics approaches provide the tools to unravel this complex system, that we call our body.

## References

Adlerberth I, Wold AE (2009) Establishment of the gut microbiota in Western infants. Acta Paediatr 98(2):229–238

Arumugam M, Raes J, Pelletier E, LE Paslier D, Yamada T, Mende DR, Fernandes GR, Tap J, Bruls T, Batto J (2011) Enterotypes of the human gut microbiome. Nature 473(7346):174–180

Azad MB, Konya T, Maughan H, Guttman DS, Field CJ, Sears MR, Becker AB, Scott JA, Kozyrskyj AL (2013) Infant gut microbiota and the hygiene hypothesis of allergic disease: impact of household pets and siblings on microbiota composition and diversity. Allergy Asthma Clin Immunol 9(1):15

Backhed F, Ding H, Wang T, Hooper LV, Koh GY, Nagy A, Semenkovich CF, Gordon JI (2004) The gut microbiota as an environmental factor that regulates fat storage. Proc Natl Acad Sci U S A 101(44):15718–15723

Backhed F, Manchester JK, Semenkovich CF, Gordon JI (2007) Mechanisms underlying the resistance to diet-induced obesity in germ-free mice. Proc Natl Acad Sci U S A 104(3):979–984

Benjdia A, Martens EC, Gordon JI, Berteau O (2011) Sulfatases and a radical S-adenosyl-L-methionine (AdoMet) enzyme are key for mucosal foraging and fitness of the prominent human gut symbiont, Bacteroides thetaiotaomicron. J Biol Chem 286(29):25973–25982

Bergstrom A, Skov TH, Bahl MI, Roager HM, Christensen LB, Ejlerskov KT, Molgaard C, Michaelsen KF, Licht TR (2014) Establishment of intestinal microbiota during early life: a longitudinal, explorative study of a large cohort of Danish infants. Appl Environ Microbiol 80(9):2889–2900

Brown CT, Davis-Richardson AG, Giongo A, Gano KA, Crabb DB, Mukherjee N, Casella G, Drew JC, Ilonen J, Knip M (2011) Gut microbiome metagenomics analysis suggests a functional model for the development of autoimmunity for type 1 diabetes. PLoS ONE 6(10):e25792

Brugman S, Klatter FA, Visser JT, Wildeboer-Veloo AC, Harmsen HJ, Rozing J, Bos NA (2006) Antibiotic treatment partially protects against type 1 diabetes in the Bio-Breeding diabetes-prone rat. Is the gut flora involved in the development of type 1 diabetes? Diabetologia 49(9):2105–2108

Cani PD, Amar J, Iglesias MA, Poggi M, Knauf C, Bastelica D, Neyrinck AM, Fava F, Tuohy KM, Chabo C, Waget A, Delmee E, Cousin B, Sulpice T, Chamontin B, Ferrieres J, Tanti JF, Gibson GR, Casteilla L, Delzenne NM, Alessi MC, Burcelin R (2007) Metabolic endotoxemia initiates obesity and insulin resistance. Diabetes 56(7):1761–1772

Chen S, Liu X, Liu J, Yang X, Lu F (2014) Ulcerative colitis as a polymicrobial infection characterized by sustained broken mucus barrier. World J Gastroenterol: WJG 20(28):9468

Claesson MJ, Cusack S, O'Sullivan O, Greene-Diniz R, DE Weerd H, Flannery E, Marchesi JR, Falush D, Dinan T, Fitzgerald G, Stanton C, VAN Sinderen D, O'Connor M, Harnedy N, O'Connor K, Henry C, O'Mahony D, Fitzgerald AP, Shanahan F, Twomey C, Hill C, Ross RP, O'Toole PW (2011) Composition, variability, and temporal stability of the intestinal microbiota of the elderly. Proc Natl Acad Sci U S A 108(Suppl 1):4586–4591

Cronise RJ, Sinclair DA, Bremer AA (2014) The "metabolic winter" hypothesis: a cause of the current epidemics of obesity and cardiometabolic disease. Metab Syndr Relat Disord 12(7):355–361

David LA, Maurice CF, Carmody RN, Gootenberg DB, Button JE, Wolfe BE, Ling AV, Devlin AS, Varma Y, Fischbach MA (2013) Diet rapidly and reproducibly alters the human gut microbiome. Nature 505(7484):559–563

DE Filippo C, Cavalieri D, DI Paola M, Ramazzotti M, Poullet JB, Massart S, Collini S, Pieraccini G, Lionetti P (2010) Impact of diet in shaping gut microbiota revealed by a comparative study in children from Europe and rural Africa. Proc Natl Acad Sci U S A 107(33):14691–14696

DE Goffau MC, Luopajarvi K, Knip M, Ilonen J, Ruohtula T, Harkonen T, Orivuori L, Hakala S, Welling GW, Harmsen HJ, Vaarala O (2013) Fecal microbiota composition differs between children with beta-cell autoimmunity and those without. Diabetes 62(4):1238–1244

DE Goffau MC, Fuentes S, VAN DEN Bogert B, Honkanen H, DE Vos WM, Welling GW, Hyöty H, Harmsen HJ (2014) Aberrant gut microbiota composition at the onset of type 1 diabetes in young children. Diabetologia 57(8):1569–1577

De Vrieze J (2014) Gut instinct. Science (New York, NY) 343(6168):241–243

Derrien M, Vaughan EE, Plugge CM, DE Vos WM (2004) Akkermansia muciniphila gen. nov., sp. nov., a human intestinal mucin-degrading bacterium. Int J Syst Evol Microbiol 54(Pt 5):1469–1476

Dominguez-Bello MG, Costello EK, Contreras M, Magris M, Hidalgo G, Fierer N, Knight R (2010) Delivery mode shapes the acquisition and structure of the initial microbiota across multiple body habitats in newborns. Proc Natl Acad Sci U S A 107(26):11971–11975

Dunne J, Triplett E, Gevers D, Xavier R, Insel R, Danska J, Atkinson M (2014) The intestinal microbiome in type 1 diabetes. Clin Exp Immunol 177:30–7

Ellis RJ, Bruce KD, Jenkins C, Stothard JR, Ajarova L, Mugisha L, Viney ME (2013) Comparison of the distal gut microbiota from people and animals in Africa. PLoS ONE 8(1):e54783

Everard A, Geurts L, Caesar R, Van Hul M, Matamoros S, Duparc T, Denis RG, Cochez P, Pierard F, Castel J (2014) Intestinal epithelial MyD88 is a sensor switching host metabolism towards obesity according to nutritional status. Nat Commun 5:5648

Fasano A (2012) Intestinal permeability and its regulation by zonulin: diagnostic and therapeutic implications. Clin Gastroenterol Hepatol 10(10):1096–1100

Favier CF, Vaughan EE, DE Vos WM, Akkermans AD (2002) Molecular monitoring of succession of bacterial communities in human neonates. Appl Environ Microbiol 68(1):219–226

Fiocchi C (1998) Inflammatory bowel disease: etiology and pathogenesis. Gastroenterology 115(1):182–205

Fleissner CK, Huebel N, Abd El-Bary MM, Loh G, Klaus S, Blaut M (2010) Absence of intestinal microbiota does not protect mice from diet-induced obesity. Br J Nutr 104(06):919–929

Flint HJ, Bayer EA, Rincon MT, Lamed R, White BA (2008) Polysaccharide utilization by gut bacteria: potential for new insights from genomic analysis. Nat Rev Microbiol 6(2):121–131

Flint HJ, Scott KP, Louis P, Duncan SH (2012a) The role of the gut microbiota in nutrition and health. Nat Rev Gastroenterol Hepatol 9(10):577–589

Flint HJ, Scott KP, Duncan SH, Louis P, Forano E (2012b) Microbial degradation of complex carbohydrates in the gut. Gut Microbes 3(4):289–306

Flint HJ, Duncan SH, Scott KP, Louis P (2014) Links between diet, gut microbiota composition and gut metabolism. Proc Nutr Soc 74:1–10

Fries W, Comunale S (2011) Ulcerative colitis: pathogenesis. Curr Drug Targets 12(10):1373–1382

Gill SR, Pop M, Deboy RT, Eckburg PB, Turnbaugh PJ, Samuel BS, Gordon JI, Relman DA, Fraser-Liggett CM, Nelson KE (2006) Metagenomic analysis of the human distal gut microbiome. Science (New York, NY) 312(5778):1355–1359

Giongo A, Gano KA, Crabb DB, Mukherjee N, Novelo LL, Casella G, Drew JC, Ilonen J, Knip M, Hyöty H (2010) Toward defining the autoimmune microbiome for type 1 diabetes. ISME J 5(1):82–91

Grönlund M, Gueimonde M, Laitinen K, Kociubinski G, Grönroos T, Salminen S, Isolauri E (2007) Maternal breast-milk and intestinal bifidobacteria guide the compositional development of the Bifidobacterium microbiota in infants at risk of allergic disease. Clin Exp Allergy 37(12):1764–1772

Hague A, Butt AJ, Paraskeva C (1996) The role of butyrate in human colonic epithelial cells: an energy source or inducer of differentiation and apoptosis? Proc Nutr Soc 55(03):937–943

Hamer HM, Jonkers DM, Bast A, Vanhoutvin SA, Fischer MA, Kodde A, Troost FJ, Venema K, Brummer RM (2009) Butyrate modulates oxidative stress in the colonic mucosa of healthy humans. Clin Nutr 28(1):88–93

Harmsen HJ, Wildeboer-Veloo AC, Raangs GC, Wagendorp AA, Klijn N, Bindels JG, Welling GW (2000) Analysis of intestinal flora development in breast-fed and formula-fed infants by using molecular identification and detection methods. J Pediatr Gastroenterol Nutr 30(1):61–67

Harmsen HJ, Raangs GC, He T, Degener JE, Welling GW (2002) Extensive set of 16S rRNA-based probes for detection of bacteria in human feces. Appl Environ Microbiol 68(6):2982–2990

Harmsen HJ, Pouwels SD, Funke A, Bos NA, Dijkstra G (2012) Crohn's disease patients have more IgG-binding fecal bacteria than controls. Clin Vaccine Immunol: CVI 19(4):515–521

Heaton KW, Radvan J, Cripps H, Mountford RA, Braddon FE, Hughes AO (1992) Defecation frequency and timing, and stool form in the general population: a prospective study. Gut 33(6):818–824

Holmes E, Li JV, Marchesi JR, Nicholson JK (2012) Gut microbiota composition and activity in relation to host metabolic phenotype and disease risk. Cell Metab 16(5):559–564

Hooper LV, Gordon JI (2001) Commensal host-bacterial relationships in the gut. Science 292(5519):1115–1118

Isolauri E (2012) Development of healthy gut microbiota early in life. J Paediatr Child Health 48(s3):1–6

Kernbauer E, Ding Y, Cadwell K (2014) An enteric virus can replace the beneficial function of commensal bacteria. Nature 516(7529):94–98

Khan MT, Nieuwdorp M, Bäckhed F (2014) Microbial modulation of insulin sensitivity. Cell Metab 20:753–760

Ley RE, Bäckhed F, Turnbaugh P, Lozupone CA, Knight RD, Gordon JI (2005) Obesity alters gut microbial ecology. Proc Natl Acad Sci U S A 102(31):11070–11075

Ley RE, Hamady M, Lozupone C, Turnbaugh PJ, Ramey RR, Bircher JS, Schlegel ML, Tucker TA, Schrenzel MD, Knight R, Gordon JI (2008) Evolution of mammals and their gut microbes. Science (New York, NY) 320(5883):1647–1651

Liverani E, Scaioli E, Cardamone C, Dal Monte P, Belluzzi A (2014) Mycobacterium avium subspecies paratuberculosis in the etiology of Crohn's disease, cause or epiphenomenon? World J Gastroenterol: WJG 20(36):13060

Lopez-Siles M, Khan TM, Duncan SH, Harmsen HJ, Garcia-Gil LJ, Flint HJ (2012) Cultured representatives of two major phylogroups of human colonic Faecalibacterium prausnitzii can utilize pectin, uronic acids, and host-derived substrates for growth. Appl Environ Microbiol 78(2):420–428

Lozupone CA, Stombaugh J, Gonzalez A, Ackermann G, Wendel D, Vazquez-Baeza Y, Jansson JK, Gordon JI, Knight R (2013) Meta-analyses of studies of the human microbiota. Genome Res 23(10):1704–1714

Martens EC, Koropatkin NM, Smith TJ, Gordon JI (2009) Complex glycan catabolism by the human gut microbiota: the Bacteroidetes Sus-like paradigm. J Biol Chem 284(37):24673–24677

Martín R, Heilig HG, Zoetendal EG, Jiménez E, Fernández L, Smidt H, Rodríguez JM (2007) Cultivation-independent assessment of the bacterial diversity of breast milk among healthy women. Res Microbiol 158(1):31–37

Martin R, Jimenez E, Heilig H, Fernandez L, Marin ML, Zoetendal EG, Rodriguez JM (2009) Isolation of bifidobacteria from breast milk and assessment of the bifidobacterial population by PCR-denaturing gradient gel electrophoresis and quantitative real-time PCR. Appl Environ Microbiol 75(4):965–969

Mejía-León ME, Petrosino JF, Ajami NJ, Domínguez-Bello MG, Calderón DLBAM (2014) Fecal microbiota

imbalance in Mexican children with type 1 diabetes. Sci Rep 4:3814
Miquel S, Martin R, Rossi O, Bermudez-Humaran LG, Chatel JM, Sokol H, Thomas M, Wells JM, Langella P (2013) Faecalibacterium prausnitzii and human intestinal health. Curr Opin Microbiol 16(3):255–261
Moore L, Moore E, Murray R, Stackebrandt E, Starr M (1987) Report of the ad hoc committee on reconciliation of approaches to bacterial systematics. Int J Syst Bacteriol 37:463–464
Morgan XC, Tickle TL, Sokol H, Gevers D, Devaney KL, Ward DV, Reyes JA, Shah SA, Leleiko N, Snapper SB (2012) Dysfunction of the intestinal microbiome in inflammatory bowel disease and treatment. Genome Biol 13(9):R79
Morotomi M, Nagai F, Watanabe Y (2012) Description of Christensenella minuta gen. nov., sp. nov., isolated from human faeces, which forms a distinct branch in the order Clostridiales, and proposal of Christensenellaceae fam. nov. Int J Syst Evol Microbiol 62(Pt 1):144–149
Murri M, Leiva I, Gomez-Zumaquero JM, Tinahones FJ, Cardona F, Soriguer F, Queipo-Ortuno MI (2013) Gut microbiota in children with type 1 diabetes differs from that in healthy children: a case-control study. BMC Med 11:46, 7015-11-46
Ouwerkerk JP, DE Vos WM, Belzer C (2013) Glycobiome: bacteria and mucus at the epithelial interface. Best Pract Res Clin Gastroenterol 27(1):25–38
Parfrey LW, Walters WA, Knight R (2011) Microbial eukaryotes in the human microbiome: ecology, evolution, and future directions. Front Microbiol 2:153
Penders J, Thijs C, Vink C, Stelma FF, Snijders B, Kummeling I, van den Brandt PA, Stobberingh EE (2006) Factors influencing the composition of the intestinal microbiota in early infancy. Pediatrics 118(2):511–521
Penders J, Thijs C, van den Brandt PA, Kummeling I, Snijders B, Stelma F, Adams H, van Ree R, Stobberingh EE (2007) Gut microbiota composition and development of atopic manifestations in infancy: the KOALA Birth Cohort Study. Gut 56(5):661–667
Peng L, Li ZR, Green RS, Holzman IR, Lin J (2009) Butyrate enhances the intestinal barrier by facilitating tight junction assembly via activation of AMP-activated protein kinase in Caco-2 cell monolayers. J Nutr 139(9):1619–1625
Qin J, Li R, Raes J, Arumugam M, Burgdorf KS, Manichanh C, Nielsen T, Pons N, Levenez F, Yamada T, Mende DR, Li J, Xu J, Li S, Li D, Cao J, Wang B, Liang H, Zheng H, Xie Y, Tap J, Lepage P, Bertalan M, Batto JM, Hansen T, Le Paslier D, Linneberg A, Nielsen HB, Pelletier E, Renault P (2010) A human gut microbial gene catalogue established by metagenomic sequencing. Nature 464:59–65
Qin J, Li Y, Cai Z, Li S, Zhu J, Zhang F, Liang S, Zhang W, Guan Y, Shen D (2012) A metagenome-wide association study of gut microbiota in type 2 diabetes. Nature 490(7418):55–60
Qin N, Yang F, Li A, Prifti E, Chen Y, Shao L, Guo J, Le Chatelier E, Yao J, Wu L (2014) Alterations of the human gut microbiome in liver cirrhosis. Nature 513(7516):59–64
Rajilić-Stojanović M, Vos WM (2014) The First 1000 cultured species of the human gastrointestinal microbiota. FEMS Microbiol Rev 38(5):996–1047
Roelofsen H, Priebe M, Vonk R (2010) The interaction of short-chain fatty acids with adipose tissue: relevance for prevention of type 2 diabetes. Benefic Microbes 1(4):433–437
Sadaghian Sadabad M, Regeling A, De Goffau MC, Blokzijl T, Weersma RK, Penders J, Faber KN, Harmsen HJ, Dijkstra G (2014) The ATG16L1-T300A allele impairs clearance of pathosymbionts in the inflamed ileal mucosa of Crohn's disease patients. Gut 64(10):1546–1552
Samuel BS, Hansen EE, Manchester JK, Coutinho PM, Henrissat B, Fulton R, Latreille P, Kim K, Wilson RK, Gordon JI (2007) Genomic and metabolic adaptations of Methanobrevibacter smithii to the human gut. Proc Natl Acad Sci U S A 104(25):10643–10648
Schwiertz A, Taras D, Schäfer K, Beijer S, Bos NA, Donus C, Hardt PD (2010) Microbiota and SCFA in lean and overweight healthy subjects. Obesity 18(1):190–195
Sekirov I, Russell SL, Antunes LC, Finlay BB (2010) Gut microbiota in health and disease. Physiol Rev 90(3):859–904
Sela DA, Chapman J, Adeuya A, Kim JH, Chen F, Whitehead TR, Lapidus A, Rokhsar DS, Lebrilla CB, German JB, Price NP, Richardson PM, Mills DA (2008) The genome sequence of Bifidobacterium longum subsp. infantis reveals adaptations for milk utilization within the infant microbiome. Proc Natl Acad Sci U S A 105(48):18964–18969
Sokol H, Pigneur B, Watterlot L, Lakhdari O, Bermudez-Humaran LG, Gratadoux JJ, Blugeon S, Bridonneau C, Furet JP, Corthier G, Grangette C, Vasquez N, Pochart P, Trugnan G, Thomas G, Blottiere HM, Dore J, Marteau P, Seksik P, Langella P (2008) Faecalibacterium prausnitzii is an anti-inflammatory commensal bacterium identified by gut microbiota analysis of Crohn disease patients. Proc Natl Acad Sci U S A 105(43):16731–16736
Sokol H, Seksik P, Furet JP, Firmesse O, Nion-Larmurier I, Beaugerie L, Cosnes J, Corthier G, Marteau P, Dore J (2009) Low counts of Faecalibacterium prausnitzii in colitis microbiota. Inflamm Bowel Dis 15:1183–1189
Spees AM, Wangdi T, Lopez CA, Kingsbury DD, Xavier MN, Winter SE, Tsolis RM, Baumler AJ (2013) Streptomycin-induced inflammation enhances Escherichia coli gut colonization through nitrate respiration. mBio 4(4):e00430-13. doi:10.1128/mBio.00430-13
Strober W (2011) Adherent-invasive E. coli in Crohn disease: bacterial "agent provocateur". J Clin Invest 121(3):841–844

Tap J, Mondot S, Levenez F, Pelletier E, Caron C, Furet J, Ugarte E, Muñoz-Tamayo R, Paslier DL, Nalin R (2009) Towards the human intestinal microbiota phylogenetic core. Environ Microbiol 11(10):2574–2584

Tims S, Derom C, Jonkers DM, Vlietinck R, Saris WH, Kleerebezem M, DE Vos WM, Zoetendal EG (2012) Microbiota conservation and BMI signatures in adult monozygotic twins. ISME J 7(4):707–717

Turnbaugh PJ, Gordon JI (2009) The core gut microbiome, energy balance and obesity. J Physiol 587:4153–4158

Turnbaugh PJ, Quince C, Faith JJ, Mchardy AC, Yatsunenko T, Niazi F, Affourtit J, Egholm M, Henrissat B, Knight R, Gordon JI (2010) Organismal, genetic, and transcriptional variation in the deeply sequenced gut microbiomes of identical twins. Proc Natl Acad Sci U S A 107(16):7503–7508

Vaarala O, Atkinson MA, Neu J (2008) The "perfect storm" for type 1 diabetes: the complex interplay between intestinal microbiota, gut permeability, and mucosal immunity. Diabetes 57(10):2555–2562

Van Nimwegen FA, Penders J, Stobberingh EE, Postma DS, Koppelman GH, Kerkhof M, Reijmerink NE, Dompeling E, Van Den Brandt PIETA, Ferreira I (2011) Mode and place of delivery, gastrointestinal microbiota, and their influence on asthma and atopy. J Allergy Clin Immunol 128(5):948–955, e3

Vignini A, Giulietti A, Nanetti L, Raffaelli F, Giusti L, Mazzanti L, Provinciali L (2013) Alzheimer's disease and diabetes: new insights and unifying therapies. Curr Diabet Rev 9(3):218–227

Vrieze A, Van Nood E, Holleman F, Salojärvi J, Kootte RS, Bartelsman JF, Dallinga–Thie GM, Ackermans MT, Serlie MJ, Oozeer R (2012) Transfer of intestinal microbiota from lean donors increases insulin sensitivity in individuals with metabolic syndrome. Gastroenterology 143(4):913–916, e7

Walker AW, Ince J, Duncan SH, Webster LM, Holtrop G, Ze X, Brown D, Stares MD, Scott P, Bergerat A (2010) Dominant and diet-responsive groups of bacteria within the human colonic microbiota. ISME J 5(2):220–230

Walter J, Ley R (2011) The human gut microbiome: ecology and recent evolutionary changes. Annu Rev Microbiol 65:411–429

Weinstock JV, Elliott DE (2009) Helminths and the IBD hygiene hypothesis. Inflamm Bowel Dis 15(1):128–133

Whitman WB, Parte AC (2009) Systematic bacteriology. Springer, New York

Willing B, Halfvarson J, Dicksved J, Rosenquist M, Järnerot G, Engstrand L, Tysk C, Jansson JK (2009) Twin studies reveal specific imbalances in the mucosa-associated microbiota of patients with ileal Crohn's disease. Inflamm Bowel Dis 15(5):653–660

Winter SE, Winter MG, Xavier MN, Thiennimitr P, Poon V, Keestra AM, Laughlin RC, Gomez G, Wu J, Lawhon SD, Popova IE, Parikh SJ, Adams LG, Tsolis RM, Stewart VJ, Baumler AJ (2013) Host-derived nitrate boosts growth of E. coli in the inflamed gut. Science (New York, NY) 339(6120):708–711

Wrzosek L, Miquel S, Noordine ML, Bouet S, Joncquel Chevalier-Curt M, Robert V, Philippe C, Bridonneau C, Cherbuy C, Robbe-Masselot C, Langella P, Thomas M (2013) Bacteroides thetaiotaomicron and Faecalibacterium prausnitzii influence the production of mucus glycans and the development of goblet cells in the colonic epithelium of a gnotobiotic model rodent. BMC Biol 11:61, 7007-11-61

Wu GD, Chen J, Hoffmann C, Bittinger K, Chen YY, Keilbaugh SA, Bewtra M, Knights D, Walters WA, Knight R, Sinha R, Gilroy E, Gupta K, Baldassano R, Nessel L, Li H, Bushman FD, Lewis JD (2011) Linking long-term dietary patterns with gut microbial enterotypes. Science (New York, NY) 334(6052):105–108

Xavier R, Podolsky D (2007) Unravelling the pathogenesis of inflammatory bowel disease. Nature 448(7152):427–434

Yatsunenko T, Rey FE, Manary MJ, Trehan I, Dominguez-Bello MG, Contreras M, Magris M, Hidalgo G, Baldassano RN, Anokhin AP (2012) Human gut microbiome viewed across age and geography. Nature 486(7402):222–227

Yutin N, Galperin MY (2013) A genomic update on clostridial phylogeny: gram-negative spore formers and other misplaced clostridia. Environ Microbiol 15(10):2631–2641

Ze X, Duncan SH, Louis P, Flint HJ (2012) Ruminococcus bromii is a keystone species for the degradation of resistant starch in the human colon. ISME J 6(8):1535–1543

Zivkovic AM, German JB, Lebrilla CB, Mills DA (2011) Human milk glycobiome and its impact on the infant gastrointestinal microbiota. Proc Natl Acad Sci U S A 108(Suppl 1):4653–4658

Zoetendal E, Akkermans A, Vliet W, Arjan J, De Visser GM, De Vos W (2001) The host genotype affects the bacterial community in the human gastronintestinal tract. Microb Ecol Health Dis 13(3):129–134

# Manipulation of the Microbiota Using Probiotics

8

Verena Grimm and Christian U. Riedel

**Abstract**

A number of diseases are associated with alterations in the composition of the microbiota of various niches of the human body. Although, in most cases, it is unclear if these alterations are the cause or the consequence of disease, they provide a rationale for therapeutic or prophylactic manipulation of a dysbiotic microbiota. Approaches to manipulate the microbiome include administration of either live bacteria, which are underrepresented in the diseased individual, substances that aim at increasing the populations of these bacteria, or a combination of the two. This chapter summarizes the available data in therapeutic manipulation of a various diseased states including irritable bowel syndrome, inflammatory bowel disease, necrotizing enterocolitis, atopic and allergic diseases, and antibiotic-associated and infectious diarrhoea.

**Keywords**

IBS • IBD • Allergy • Necrotising enterocolitis • Diarrhea

## 8.1 Introduction

The various epithelial interfaces of the human body with the outside environment harbour unique and characteristic microbial communities that are specific for an individual and relatively stable over time under normal circumstances (Human Microbiome Project Consortium 2012; Faith et al. 2013; Grice and Segre 2011; Gajer et al. 2012). Nevertheless, characteristic changes in the composition of the microbiota have been observed in a number of diseases including inflammatory bowel disease (IBD), irritable bowel syndrome (IBS), atopic and allergic, periodontitis, infectious diseases, metabolic syndrome, and cancer (Sears and Garrett 2014; Schwabe and Jobin 2013; Sekirov et al. 2010; Gerritsen et al. 2011; Sommer and Bäckhed 2013; Russell and Finlay 2012). Interestingly, dysbioses in the microbiota have also been

V. Grimm (✉) • C.U. Riedel
Institute of Microbiology and Biotechnology, University of Ulm,
Albert-Einstein-Allee 11, 89069 Ulm, Germany
e-mail: verena.grimm@uni-ulm.de; christian.riedel@uni-ulm.de



A. Schwiertz (ed.), *Microbiota of the Human Body*, Advances in Experimental Medicine and Biology 902, DOI 10.1007/978-3-319-31248-4_8

observed in psychiatric disorders including autism and depression (Sekirov et al. 2010; Dinan and Cryan 2013). The majority of studies link changes in the intestinal microbiota with disease (Sekirov et al. 2010; Sommer and Bäckhed 2013) with less but increasing data available on disease-related changes in the vaginal (Ma et al. 2012; Brotman 2011), oral (Pihlstrom et al. 2005; Wang et al. 2013), or skin (Grice and Segre 2011; Zeeuwen et al. 2013) microbiota.

In most cases, the question whether an altered microbiota is the cause or the consequence of disease is not definitively answered. Nevertheless, these changes are the rationale for therapeutic or prophylactic interventions attempting to manipulate or restore the microbiota for which a dysbiosis has been observed. The approaches to manipulate the microbiota for therapeutic reasons include the administration of either (potentially) beneficial microorganisms (probiotics), substrates to boost specific populations of microorganisms (prebiotics), or a combination of both (synbiotics).

The term "probiotic" was initially coined by Lilly and Stillwell describing an unidentified factor produced by one ciliate (*Colpidium campylum*) that promoted growth of another (*Tetrahymena pyriformis*) (Lilly and Stillwell 1965). Today, probiotics are defined as 'live microorganisms which when administered in adequate amounts confer a health benefit on the host' according to guidelines set by the World Health Organisation and the Food and Agriculture Organization of the United Nations (FAO/WHO Working Group 2002). Most of the organisms used and marketed as probiotics belong to the genera *Lactobacillus* and *Bifidobacterium* with a few strains of other lactic acid bacteria, *Bacillus sp.*, *Saccharomyces sp.*, and *E. coli* (Gareau et al. 2010; Bron et al. 2012; Foligné et al. 2013). Probiotic bacteria are marketed and administered as naturally fermented or non-fermented food products in various matrices, food supplement powders, tablets etc. and these products either contain single strains or mixes of several probiotic bacteria (Gareau et al. 2010; Foligné et al. 2013).

The analysis of the composition of the microbiota in diverse habitats of the human body as well as therapeutic approaches using prebiotics, synbiotics or microbiota transplantation will be discussed elsewhere in this book. In this chapter, we will focus the clinical data available for probiotics treatment in a number of important diseases and, where appropriate, we will discuss alterations in the microbiota that are reported for these diseases. Due to the large number of clinical trials performed on probiotic treatments, we will focus on available review articles and meta-analyses on the efficacy of probiotic treatments. Since the microorganisms used as probiotics generally belong to the group of GRAS organisms, we will not elude to studies, reviews, or meta-analyses dealing with the safety aspects of probiotic treatments. We will not discuss individual probiotics or formulations in a specific disease. Instead, general trends in the efficacy of probiotics in these diseases found in the meta-analyses will be summarized. For results on specific probiotics, the reader is referred to individual studies included in these meta-analyses.

## 8.2 Irritable Bowel Syndrome

IBS is a group of complex intestinal disorders that are characterized by a range of very diverse, sometimes contradictory symptoms including chronic abdominal pain, bloating and altered stool frequency (constipation, diarrhoea, or both with alternating episodes). Due to the lack of validated genetic, biochemical or physiological markers, diagnosis of IBS is based on intensive examination and exclusion of other diseases. About 10–20 % of adults and adolescents display symptoms of IBS with a dominance in female patients. Affected persons are classified into different patients subgroups according to the ROME criteria based their symptoms (Longstreth et al. 2006; Chang and Talley 2011). The pathogenesis of IBS is multifactorial but there is strong evidence for a contribution of the GIT microbiota. Changes in the composition of both the faecal and the mucosal microbiota have been observed and these changes appear to be specific for IBS subgroups (Gerritsen et al. 2011; Jeffery et al. 2012; Kerckhoffs et al. 2009; Rajilić-Stojanović et al. 2011). However, a unifying feature of all IBS subtypes seems to be a reduction in bifidobacteria (Kerckhoffs et al. 2009; Rajilić-Stojanović et al. 2011; Rigsbee et al. 2012).

Given the (likely) contribution of the microbiota in pathogenesis and the reduced abundance of bifidobacteria in the GIT of patients, probiotics represent an logical therapeutic intervention of IBS. In recent years, a number of reviews and meta-analyses have summarized the clinical data available for probiotic treatments in IBS (Hoveyda et al. 2009; McFarland and Dublin 2008; Moayyedi et al. 2010; Brenner et al. 2009; Whelan 2011; Whelan and Quigley 2013; Ritchie and Romanuk 2012). These analyses included a total of 30 clinical trials, in which patients were treated with various probiotics containing different strains of bifidobacteria, lactobacilli, streptococci, *Saccharomyces boulardii*, and *E. coli* as single strain preparations, probiotic mixes or synbiotics. While some of the trials were criticised for suboptimal study design, the overall outcome was that most probiotics alleviate the symptoms of IBS with some strains being more effective than others.

## 8.3 Inflammatory Bowel Disease (IBD)

Inflammatory bowel disease (IBD) is a group of chronic gastrointestinal disorders characterized by relapsing and remitting inflammation of the GIT (Bouma and Strober 2003; Cho 2008; Xavier and Podolsky 2007). The two by far most widespread forms of IBDs are ulcerative colitis (UC) and Crohn's disease (CD). Both UC and CD share most of the symptoms but differ in the extent and anatomic location of the inflamed tissues. UC is limited to the mucosa of the large intestine and extends from the distal to the proximal colon in varying degrees. One form of UC is pouchitis, which represents a chronic inflammation of the artificial rectum created by surgical formation of a pouch from ileal tissue in patients that have undergone colectomy. By contrast, in CD inflammation is transmural, can affect any site of the GIT, and healthy and inflamed sites can alternate.

IBDs are multifactorial diseases with a considerable contribution of genetic predisposition, environmental factors and the intestinal microbiota involved. In both UC and CD, prominent, a reduced diversity and changes the abundance of prominent bacterial groups of the microbiota have been observed (Gerritsen et al. 2011; Peterson et al. 2008; Kostic et al. 2014). Principal component analysis shows that the microbiota profiles of UC, CD and healthy controls are clearly different from each other (Qin et al. 2010).

For CD, extensive 16S rRNA gene sequencing of faecal and mucosal samples obtained from 447 children with on-set, untreated CD and 221 healthy controls revealed an overall reduction in diversity in diseased individuals (Gevers et al. 2014). Moreover, an increase in *Enterobacteriaceae*, *Pasteurellaceae*, *Veillonellaceae*, and *Fusobacteriaceae*, and a reduction in *Erysipelotrichaceae*, *Bacterdoidales*, and *Clostridiales* were observed in mucosal but not faecal samples. This suggests that changes in the mucosal microbiota are more prominent and important for development of CD than changes in the composition of the luminal microbiota. In other studies similar observations were made with an increase in *Proteobacteria* and a reduction in *Firmicutes* and *Bacteroidetes* and potentially beneficial or anti-inflammatory bacteria including *Faecalibacterium prausnitzii* or bifidobacteria (Kostic et al. 2014; Franks et al. 1998; Sokol et al. 2008; Schwiertz et al. 2010; Joossens et al. 2011).

Meta analyses on the effect on probiotics in UC and pouchitis uniformly suggest that probiotics are more effective than placebo in maintaining remission (Shen et al. 2014; Holubar et al. 2010; Jin et al. 2007). However, probiotics did not show a positive effect on recurrence in post-operative CD compared to placebo (Doherty et al. 2009, 2010; Van Loo et al. 2012). Nevertheless, it was noted that the effect of probiotics in the management of CD merits further investigation (Doherty et al. 2009, 2010; Van Loo et al. 2012). In the study by Gevers et al. (2014), in a subgroup of 57 diseased children that already had received antibiotics at the time point of sampling the dysbiosis was more pronounced compared to the other, non-treated samples. This indicates that use of antibiotics for treatment of CD might actually increase dysbiosis questioning the use of antibiotics for treatment of CD. Thus, in light of problems associated with antibotic use in increasing dysbiosis and the recommendations of the meta analyses, probiotics might become an alternative strategy in the management of CD but need further investigation.

## 8.4 Atopy/Allergy

Atopy and allergic diseases such as asthma, hay fever, food allergy, and atopic dermatitis (AD) are hypersensitivity disorders caused by an aberrant immune response to environmental antigens (Kay 2000). Affected persons react to normally harmless substances abundant in the environment such as food antigens, pollen, and commensal bacteria. The aberrant immune response in allergic patients is characterized by a predominant Th2 response resulting in excessive activation of mast cells, basophils, and IgE producing B cells. The symptoms of allergy are the result of the release of local pro-inflammatory mediators including histamines from these cells. Standard therapies include treatment with antihistamines in the acute phase and hypersensitisation for prophylaxis.

Since atopic and allergic diseases have seen an enormous increase in the last five decades coinciding with improved sanitary and hygiene standards one of the possible explanations for this increase is the so-called hygiene hypothesis (Brooks et al. 2013). The hygiene hypothesis states that the aberrant immune responses to environmental antigens in allergy is the result of an insufficient exposure to antigen early in life. Recent data particularly on allergic asthma has raised scepticism and thus it has been suggested to revise and generalize the hygiene hypothesis (Brooks et al. 2013; Wills-Karp et al. 2001). Nevertheless, the microbiota plays a crucial role in the development of atopic diseases as demonstrated by profound shifts in the composition of the GIT and skin microbiota observed in allergic patients (Grice and Segre 2011; Russell and Finlay 2012; Zeeuwen et al. 2013). Also, the development of the microbiota is heavily influenced by the mode of delivery and early infant feeding (Matamoros et al. 2013). Moreover, caesarean section and bottle feeding have been associated with an increased risk to develop atopic diseases (Bager et al. 2008; Yang et al. 2009).

In consequence, probiotics have come into focus as an alternative or supplementary therapeutic strategy in the management of atopic diseases. Several meta analyses have studied the impact of probiotic administration to prevent of treat allergies. Two analysis on prevention of asthma and wheeze concluded that there is not sufficient evidence to recommend administration probiotics (neither prenatally to mothers nor postnatally to infants) as a protective treatment (Azad et al. 2013; Elazab et al. 2013). However, one of the studies reported an association between probiotics and a reduced risk for sensitisation and decreased serum levels of IgE (Elazab et al. 2013). The authors of both studies suggested to follow-up on existing trials and further clinical and basic research on specific probiotic strains in childhood asthma. Interestingly, probiotics seem to be more effective for prevention of AD. Of three independent meta analysis only one concluded that probiotics can not be recommended as a treatment option to prevent AD but indicated that the results could be biased by the high heterogeneity in the probiotic strains and formulations and specific probiotics may still be protective (Boyle et al. 2009). Two more recent analyses indicated that prenatal administration of probiotics to mothers during pregnancy as well as postnatal to infants reduced the risk for AD in both the general population and at-risk groups (Doege et al. 2012; Panduru et al. 2014).

## 8.5 Necrotizing Enterocolitis

Necrotizing enterocolitis (NEC) is an acute inflammatory disorder of the intestinal mucosa and one of the leading causes for neonatal morbidity and mortality (Neu and Walker 2011; Gephart et al. 2012). While NEC primarily affects low birth-weight (less than 1,500 g), pre-term infants it occasionally also occurs in full-term infants. Symptoms of NEC include feeding intolerance, bloody stool, abdominal distention and heavy inflammation and necrosis of the intestinal tissue starting at around 8–10 days post partum.

NEC is thought to be mainly the result of an immature and highly immunoreactive intestinal mucosa (Neu and Walker 2011; Gephart et al. 2012). Reduced expression of mucins, immunoglobulin A, and tight junction proteins by the immature intestinal epithelium result in a leaky barrier and an increased exposure of cells of the underlying mucosal immune system to microbial antigens. Additionally, an increased expression in Toll-like receptor 4 and a reduced expression of the inhibitor of NF-κB suggests abnormal sensing and signalling of lipopolysaccharide (LPS), a molecule abundant in the outer membrane of Gram-negative bacteria with potent pro-inflammatory activity. Furthermore, there is strong evidence for an altered composition of the intestinal microbiota in NEC patients compared to full-term, vaginally delivered, breast-fed infants (Carlisle and Morowitz 2013; Grishin et al. 2013). The most commonly found differences include a reduced diversity and an over-abundance of *Proteobacteria*, a group of LPS-containing, Gram-negative bacteria. Collectively these factors lead to an excessive inflammatory response ultimately causing necrosis of the intestinal tissue.

The treatment options for diagnosed NEC include bowel decompression, discontinued enteral feeding, intravenous antibiotics, and surgical intervention. Since abnormal LPS/TLR4 signalling, altered microbial colonisation and other factors strongly suggest a role of the intestinal microbiota in the pathology, probiotics have gained considerable interest in the management of NEC.

In several independent meta-analyses, probiotics were found to have beneficial effects in NEC patients (Mihatsch et al. 2012; Deshpande et al. 2010; Alfaleh and Anabrees 2014). One of these analyses stated that some probiotics may reduce the severity and mortality of NEC (Mihatsch et al. 2012). By contrast, the other two analyses were more enthusiastic noting significant effects of probiotics containing lactobacilli alone or in combination with bifidobacteria in the prevention or treatment of NEC (Deshpande et al. 2010; Alfaleh and Anabrees 2014). Thus, the available data supports the use of probiotics in pre-term infants to prevent NEC or reduce the severity and mortality of the disease. In the future, further comparative studies should be performed to identify the most effective probiotic preparations and therapeutic conditions in terms of dose, timing, and duration.

## 8.6 Diarrhoeal Diseases

Globally, there are approximately 1.7 billion diarrhoeal episodes per year accounting for almost 760,000 deaths of children under 5 years of age. Diarrhoeal diseases are thus the second leading cause of childhood mortality (WHO fact sheet No. 330, April 2013). Diarrhoea is defined as three or more loose or watery stools per day. In case of acute infectious diarrhoea, the diseases is caused by a (diagnosed) infectious agent including viruses, bacteria and eukaryotic parasites (Thielman and Guerrant 2004). The vast majority of these infections are the result of consumption of water contaminated with human or animal faeces or poor hygiene standards and a shortage in safe cooking resulting in contaminated food products (WHO fact sheet No. 330, April 2013). The pathogens most frequently found to cause diarrhoea are rotavirus, norovirus, *Salmonella sp.*, *Shigella sp.*, *Campylobacter jejuni*, pathogenic *E. coli* strains, *Cryptosporidium sp.*, and *Giardia sp.* (Thielman and Guerrant 2004; DuPont 2014). Traveller's diarrhoea is a form of infectious diarrhoea and a frequent problem in persons travelling from geographic zones of low risk (Europe, USA, Canada, Japan, and Australasia) to areas with high risk (northern Africa, Latin America, Middle East and Southeast Asia) (Al-Abri et al. 2005). Another form of infectious diarrhoea is observed in 5–25 % of patients treated with a range of antibiotics (Bartlett 2002). While *C. difficile* is responsible for only 15–20 % of the cases, it is the most severe form of antibiotic-associated diarrhoea since infections often become chronic and it is responsible for almost all cases of colitis associated with antibiotic treatment (Bartlett 2002).

The normal human intestinal microbiota confers colonisation resistance against a number of important diarrhoea-causing human pathogens including *C. difficile*, *S.* Typhimurium, and others (Buffie and Pamer 2013; Ubeda and Pamer 2012; Stecher and Hardt 2008). It is thus reasonable to hypothesize that antibiotic-associated diarrhoea is the results of a disruption of the normal (intestinal) microbiota as a consequence of administration of antibiotics accompanied by rapid growth of opportunistic pathogens. In fact, this hypothesis is supported by recent metagenomic studies showing rapid and only partially reversible changes in the composition of the intestinal microbiota following antibiotic treatment (Dethlefsen and Relman 2011; Pérez-Cobas et al. 2013). Additionally, the inflammatory processs and diarrhoea alters the intestinal environment dramatically favouring growth of diarrhoeal pathogens even in the absence of antibiotics (Buffie and Pamer 2013; Stecher and Hardt 2008).

Since diarrhoeal diseases are caused by microbial pathogens, in most cases the first treatment option in normal, immunocompetent individuals are antibiotics with duration and choice of antibiotic depending on the causative agent (DuPont 2014). An alternative or supplementary treatment option is the administration of probiotics to prevent alterations of the microbiota or to more rapidly restore the normal composition. Several meta-analyses have reviewed the available clinical data on the treatment of infectious diarrhoea using probiotics (Johnston et al. 2011, 2012; Hempel et al. 2012; Videlock and Cremonini 2012; Goldenberg et al. 2013; Johnson et al. 2012). The overall results indicated that probiotics are protective in preventing antibiotic-associated diarrhoea. Moreover, probiotics reduce the stool frequency and shorten the duration of infectious diarrhoea. However, further clinical trials are needed to identify the most effective probiotic strains, doses, and treatment regimes.

## References


Al-Abri SS, Beeching NJ, Nye FJ (2005) Traveller's diarrhoea. Lancet Infect Dis 5:349–360

Alfaleh K, Anabrees J (2014) Probiotics for prevention of necrotizing enterocolitis in preterm infants. Cochrane Database Syst Rev 4:CD005496

Azad MB, Coneys JG, Kozyrskyj AL, Field CJ, Ramsey CD, Becker AB, Friesen C, Abou-Setta AM, Zarychanski R (2013) Probiotic supplementation during pregnancy or infancy for the prevention of asthma and wheeze: systematic review and meta-analysis. BMJ 347:f6471

Bager P, Wohlfahrt J, Westergaard T (2008) Caesarean delivery and risk of atopy and allergic disease: meta-analyses. Clin Exp Allergy 38:634–642

Bartlett JG (2002) Clinical practice. Antibiotic-associated diarrhea. N Engl J Med 346:334–339

Bouma G, Strober W (2003) The immunological and genetic basis of inflammatory bowel disease. Nat Rev Immunol 3:521–533

Boyle RJ, Bath-Hextall FJ, Leonardi-Bee J, Murrell DF, Tang ML-K (2009) Probiotics for the treatment of eczema: a systematic review. Clin Exp Allergy 39:1117–1127

Brenner DM, Moeller MJ, Chey WD, Schoenfeld PS (2009) The utility of probiotics in the treatment of irritable bowel syndrome: a systematic review. Am J Gastroenterol 104:1033–1049, quiz 1050

Bron PA, van Baarlen P, Kleerebezem M (2012) Emerging molecular insights into the interaction between probiotics and the host intestinal mucosa. Nat Rev Microbiol 10:66–78

Brooks C, Pearce N, Douwes J (2013) The hygiene hypothesis in allergy and asthma: an update. Curr Opin Allergy Clin Immunol 13:70–77

Brotman RM (2011) Vaginal microbiome and sexually transmitted infections: an epidemiologic perspective. J Clin Invest 121:4610–4617

Buffie CG, Pamer EG (2013) Microbiota-mediated colonization resistance against intestinal pathogens. Nat Rev Immunol 13:790–801

Carlisle EM, Morowitz MJ (2013) The intestinal microbiome and necrotizing enterocolitis. Curr Opin Pediatr 25:382–387

Chang JY, Talley NJ (2011) An update on irritable bowel syndrome: from diagnosis to emerging therapies. Curr Opin Gastroenterol 27:72–78

Cho JH (2008) The genetics and immunopathogenesis of inflammatory bowel disease. Nat Rev Immunol 8:458–466

Deshpande G, Rao S, Patole S, Bulsara M (2010) Updated meta-analysis of probiotics for preventing necrotizing enterocolitis in preterm neonates. Pediatrics 125:921–930

Dethlefsen L, Relman DA (2011) Incomplete recovery and individualized responses of the human distal gut microbiota to repeated antibiotic perturbation. Proc Natl Acad Sci U S A 108(Suppl):4554–4561

Dinan TG, Cryan JF (2013) Melancholic microbes: a link between gut microbiota and depression? Neurogastroenterol Motil 25:713–719

Doege K, Grajecki D, Zyriax B-C, Detinkina E, Zu Eulenburg C, Buhling KJ (2012) Impact of maternal supplementation with probiotics during pregnancy on atopic eczema in childhood – a meta-analysis. Br J Nutr 107:1–6

Doherty G, Bennett G, Patil S, Cheifetz A, Moss AC (2009) Interventions for prevention of post-operative recurrence of Crohn's disease. Cochrane Database Syst Rev 7:CD006873

Doherty GA, Bennett GC, Cheifetz AS, Moss AC (2010) Meta-analysis: targeting the intestinal microbiota in prophylaxis for post-operative Crohn's disease. Aliment Pharmacol Ther 31:802–809

DuPont HL (2014) Acute infectious diarrhea in immunocompetent adults. N Engl J Med 370:1532–1540

Elazab N, Mendy A, Gasana J, Vieira ER, Quizon A, Forno E (2013) Probiotic administration in early life, atopy, and asthma: a meta-analysis of clinical trials. Pediatrics 132:e666–e676

Faith JJ, Guruge JL, Charbonneau M, Subramanian S, Seedorf H, Goodman AL, Clemente JC, Knight R, Heath AC, Leibel RL, Rosenbaum M, Gordon JI (2013) The long-term stability of the human gut microbiota. Science 341:1237439

FAO/WHO Working Group on Drafting Guidelines for the Evaluation of Probiotics (2002) Guidelines for the evaluation of probiotics in food

Foligné B, Daniel C, Pot B (2013) Probiotics from research to market: the possibilities, risks and challenges. Curr Opin Microbiol 16:284–292

Franks AH, Harmsen HJ, Raangs GC, Jansen GJ, Schut F, Welling GW (1998) Variations of bacterial populations in human feces measured by fluorescent in situ hybridization with group-specific 16S rRNA-targeted oligonucleotide probes. Appl Environ Microbiol 64:3336–3345

Gajer P, Brotman RM, Bai G, Sakamoto J, Schütte UME, Zhong X, Koenig SSK, Fu L, Ma ZS, Zhou X, Abdo Z, Forney LJ, Ravel J (2012) Temporal dynamics of the human vaginal microbiota. Sci Transl Med 4:132ra52

Gareau MG, Sherman PM, Walker WA (2010) Probiotics and the gut microbiota in intestinal health and disease. Nat Rev Gastroenterol Hepatol 7:503–514

Gephart SM, McGrath JM, Effken JA, Halpern MD (2012) Necrotizing enterocolitis risk: state of the science. Adv Neonatal Care 12:77–87, quiz 88–9

Gerritsen J, Smidt H, Rijkers GT, de Vos WM (2011) Intestinal microbiota in human health and disease: the impact of probiotics. Genes Nutr 6:209–240

Gevers D, Kugathasan S, Denson LA, Vázquez-Baeza Y, Van Treuren W, Ren B, Schwager E, Knights D, Song SJ, Yassour M, Morgan XC, Kostic AD, Luo C, González A, McDonald D, Haberman Y, Walters T, Baker S, Rosh J, Stephens M, Heyman M, Markowitz J, Baldassano R, Griffiths A, Sylvester F, Mack D, Kim S, Crandall W, Hyams J, Huttenhower C, Knight R, Xavier RJ (2014) The treatment-naive microbiome in new-onset Crohn's disease. Cell Host Microbe 15:382–392

Goldenberg JZ, Ma SSY, Saxton JD, Martzen MR, Vandvik PO, Thorlund K, Guyatt GH, Johnston BC (2013) Probiotics for the prevention of Clostridium difficile-associated diarrhea in adults and children. Cochrane Database Syst Rev 5:CD006095

Grice EA, Segre JA (2011) The skin microbiome. Nat Rev Microbiol 9:244–253

Grishin A, Papillon S, Bell B, Wang J, Ford HR (2013) The role of the intestinal microbiota in the pathogenesis of necrotizing enterocolitis. Semin Pediatr Surg 22:69–75

Hempel S, Newberry SJ, Maher AR, Wang Z, Miles JNV, Shanman R, Johnsen B, Shekelle PG (2012) Probiotics for the prevention and treatment of antibiotic-associated diarrhea: a systematic review and meta-analysis. JAMA 307:1959–1969

Holubar SD, Cima RR, Sandborn WJ, Pardi DS (2010) Treatment and prevention of pouchitis after ileal pouch-anal anastomosis for chronic ulcerative colitis. Cochrane Database Syst Rev 16:CD001176

Hoveyda N, Heneghan C, Mahtani KR, Perera R, Roberts N, Glasziou P (2009) A systematic review and meta-analysis: probiotics in the treatment of irritable bowel syndrome. BMC Gastroenterol 9:15

Human Microbiome Project Consortium (2012) Structure, function and diversity of the healthy human microbiome. Nature 486:207–214

Jeffery IB, O'Toole PW, Öhman L, Claesson MJ, Deane J, Quigley EMM, Simrén M (2012) An irritable bowel syndrome subtype defined by species-specific alterations in faecal microbiota. Gut 61:997–1006

Jin MS, Kim SE, Heo JY, Lee ME, Kim HM, Paik S-G, Lee H, Lee J-O (2007) Crystal structure of the TLR1-TLR2 heterodimer induced by binding of a tri-acylated lipopeptide. Cell 130:1071–1082

Johnson S, Maziade P-J, McFarland LV, Trick W, Donskey C, Currie B, Low DE, Goldstein EJC (2012) Is primary prevention of Clostridium difficile infection pos-

sible with specific probiotics? Int J Infect Dis 16:e786–e792
Johnston BC, Goldenberg JZ, Vandvik PO, Sun X, Guyatt GH (2011) Probiotics for the prevention of pediatric antibiotic-associated diarrhea. Cochrane Database Syst Rev 29(2):232–42 CD004827
Johnston BC, Ma SSY, Goldenberg JZ, Thorlund K, Vandvik PO, Loeb M, Guyatt GH (2012) Probiotics for the prevention of Clostridium difficile-associated diarrhea: a systematic review and meta-analysis. Ann Intern Med 157:878–888
Joossens M, Huys G, Cnockaert M, De Preter V, Verbeke K, Rutgeerts P, Vandamme P, Vermeire S (2011) Dysbiosis of the faecal microbiota in patients with Crohn's disease and their unaffected relatives. Gut 60:631–637
Kay AB (2000) Overview of "allergy and allergic diseases: with a view to the future". Br Med Bull 56:843–864
Kerckhoffs APM, Samsom M, van der Rest ME, de Vogel J, Knol J, Ben-Amor K, Akkermans LMA (2009) Lower Bifidobacteria counts in both duodenal mucosa-associated and fecal microbiota in irritable bowel syndrome patients. World J Gastroenterol WJG 15:2887–2892
Kostic AD, Xavier RJ, Gevers D (2014) The microbiome in inflammatory bowel disease: current status and the future ahead. Gastroenterology 146:1489–1499
Lilly DM, Stillwell RH (1965) Probiotics: growth-promoting factors produced by microorganisms. Science 147:747–748
Longstreth GF, Thompson WG, Chey WD, Houghton LA, Mearin F, Spiller RC (2006) Functional bowel disorders. Gastroenterology 130:1480–1491
Ma B, Forney LJ, Ravel J (2012) Vaginal microbiome: rethinking health and disease. Annu Rev Microbiol 66:371–389
Matamoros S, Gras-Leguen C, Le Vacon F, Potel G, de La Cochetiere M-F (2013) Development of intestinal microbiota in infants and its impact on health. Trends Microbiol 21:167–173
McFarland LV, Dublin S (2008) Meta-analysis of probiotics for the treatment of irritable bowel syndrome. World J Gastroenterol 14:2650–2661
Mihatsch WA, Braegger CP, Decsi T, Kolacek S, Lanzinger H, Mayer B, Moreno LA, Pohlandt F, Puntis J, Shamir R, Stadtmüller U, Szajewska H, Turck D, van Goudoever JB (2012) Critical systematic review of the level of evidence for routine use of probiotics for reduction of mortality and prevention of necrotizing enterocolitis and sepsis in preterm infants. Clin Nutr 31:6–15
Moayyedi P, Ford AC, Talley NJ, Cremonini F, Foxx-Orenstein AE, Brandt LJ, Quigley EMM (2010) The efficacy of probiotics in the treatment of irritable bowel syndrome: a systematic review. Gut 59:325–332
Neu J, Walker WA (2011) Necrotizing enterocolitis. N Engl J Med 364:255–264
Panduru M, Panduru NM, Sălăvăstru CM, Tiplica G-S (2014) Probiotics and primary prevention of atopic dermatitis: a meta-analysis of randomized controlled studies. J Eur Acad Dermatol Venereol 29:232–42
Pérez-Cobas AE, Artacho A, Knecht H, Ferrús ML, Friedrichs A, Ott SJ, Moya A, Latorre A, Gosalbes MJ (2013) Differential effects of antibiotic therapy on the structure and function of human gut microbiota. PLoS One 8:e80201
Peterson DA, Frank DN, Pace NR, Gordon JI (2008) Metagenomic approaches for defining the pathogenesis of inflammatory bowel diseases. Cell Host Microbe 3:417–427
Pihlstrom BL, Michalowicz BS, Johnson NW (2005) Periodontal diseases. Lancet 366:1809–1820
Qin J, Li R, Raes J, Arumugam M, Burgdorf KS, Manichanh C, Nielsen T, Pons N, Levenez F, Yamada T, Mende DR, Li J, Xu J, Li S, Li D, Cao J, Wang B, Liang H, Zheng H, Xie Y, Tap J, Lepage P, Bertalan M, Batto J-M, Hansen T, Le Paslier D, Linneberg A, Nielsen HB, Pelletier E, Renault P, Sicheritz-Ponten T, Turner K, Zhu H, Yu C, Li S, Jian M, Zhou Y, Li Y, Zhang X, Li S, Qin N, Yang H, Wang J, Brunak S, Doré J, Guarner F, Kristiansen K, Pedersen O, Parkhill J, Weissenbach J, Bork P, Ehrlich SD, Wang J (2010) A human gut microbial gene catalogue established by metagenomic sequencing. Nature 464:59–65
Rajilić-Stojanović M, Biagi E, Heilig HGHJ, Kajander K, Kekkonen RA, Tims S, de Vos WM (2011) Global and deep molecular analysis of microbiota signatures in fecal samples from patients with irritable bowel syndrome. Gastroenterology 141:1792–1801
Rigsbee L, Agans R, Shankar V, Kenche H, Khamis HJ, Michail S, Paliy O (2012) Quantitative profiling of gut microbiota of children with diarrhea-predominant irritable bowel syndrome. Am J Gastroenterol 107:1740–1751
Ritchie ML, Romanuk TN (2012) A meta-analysis of probiotic efficacy for gastrointestinal diseases. PLoS One 7:e34938
Russell SL, Finlay BB (2012) The impact of gut microbes in allergic diseases. Curr Opin Gastroenterol 28:563–569
Sang L-X, Chang B, Zhang W-L, Wu X-M, Li X-H, Jiang M (2010) Remission induction and maintenance effect of probiotics on ulcerative colitis: a meta-analysis. World J Gastroenterol 16:1908–1915
Schwabe RF, Jobin C (2013) The microbiome and cancer. Nat Rev Cancer 13:800–812
Schwiertz A, Jacobi M, Frick J-S, Richter M, Rusch K, Köhler H (2010) Microbiota in pediatric inflammatory bowel disease. J Pediatr 157:240–244.e1
Sears CL, Garrett WS (2014) Microbes, microbiota, and colon cancer. Cell Host Microbe 15:317–328
Sekirov I, Russell SL, Antunes LCM, Finlay BB (2010) Gut microbiota in health and disease. Physiol Rev 90:859–904
Shen J, Zuo Z-X, Mao A-P (2014) Effect of probiotics on inducing remission and maintaining therapy in ulcer-

ative colitis, Crohn's disease, and pouchitis: meta-analysis of randomized controlled trials. Inflamm Bowel Dis 20:21–35
Sokol H, Pigneur B, Watterlot L, Lakhdari O, Bermúdez-Humarán LG, Gratadoux J-J, Blugeon S, Bridonneau C, Furet J-P, Corthier G, Grangette C, Vasquez N, Pochart P, Trugnan G, Thomas G, Blottière HM, Doré J, Marteau P, Seksik P, Langella P (2008) Faecalibacterium prausnitzii is an anti-inflammatory commensal bacterium identified by gut microbiota analysis of Crohn disease patients. Proc Natl Acad Sci U S A 105:16731–16736
Sommer F, Bäckhed F (2013) The gut microbiota – masters of host development and physiology. Nat Rev Microbiol 11:227–238
Stecher B, Hardt W-D (2008) The role of microbiota in infectious disease. Trends Microbiol 16:107–114
Thielman NM, Guerrant RL (2004) Clinical practice. Acute infectious diarrhea. N Engl J Med 350:38–47
Ubeda C, Pamer EG (2012) Antibiotics, microbiota, and immune defense. Trends Immunol 33:459–466
Van Loo ES, Dijkstra G, Ploeg RJ, Nieuwenhuijs VB (2012) Prevention of postoperative recurrence of Crohn's disease. J Crohns Colitis 6:637–646
Videlock EJ, Cremonini F (2012) Meta-analysis: probiotics in antibiotic-associated diarrhoea. Aliment Pharmacol Ther 35:1355–1369
Wang J, Qi J, Zhao H, He S, Zhang Y, Wei S, Zhao F (2013) Metagenomic sequencing reveals microbiota and its functional potential associated with periodontal disease. Sci Rep 3:1843
Whelan K (2011) Probiotics and prebiotics in the management of irritable bowel syndrome: a review of recent clinical trials and systematic reviews. Curr Opin Clin Nutr Metab Care 14:581–587
Whelan K, Quigley EMM (2013) Probiotics in the management of irritable bowel syndrome and inflammatory bowel disease. Curr Opin Gastroenterol 29:184–189
WHO (Fact Sheet No. 330, April 2013) Diarrhoeal disease. World Health Organization. http://www.who.int/mediacentre/factsheets/fs330/en/
Wills-Karp M, Santeliz J, Karp CL (2001) The germless theory of allergic disease: revisiting the hygiene hypothesis. Nat Rev Immunol 1:69–75
Xavier RJ, Podolsky DK (2007) Unravelling the pathogenesis of inflammatory bowel disease. Nature 448:427–434
Yang YW, Tsai CL, Lu CY (2009) Exclusive breastfeeding and incident atopic dermatitis in childhood: a systematic review and meta-analysis of prospective cohort studies. Br J Dermatol 161:373–383
Zeeuwen PLJM, Kleerebezem M, Timmerman HM, Schalkwijk J (2013) Microbiome and skin diseases. Curr Opin Allergy Clin Immunol 13:514–520

# How to Manipulate the Microbiota: Prebiotics

# 9

Petra Louis, Harry J. Flint, and Catherine Michel

**Abstract**

During the last century, human nutrition has evolved from the definition of our nutritional needs and the identification of ways to meet them, to the identification of food components that can optimise our physiological and psychological functions. This development, which aims to ensure the welfare, health and reduced susceptibility to disease during life, gave birth to the concept of "functional foods". In this context, there is an increasing interest in the physiological effects induced by the dense and diverse microbiota which inhabits the human colon and whose development depends on the fermentation of undigested food residues. Thus, much research aims at identifying ways to guide these impacts in order to benefit the health of the host. It is in this context that the concept of "prebiotics" was developed in the 1990s. Since then, prebiotics have stimulated extensive work in order to clarify their definition, their nature and their physiological properties in accordance with the evolution of knowledge on the intestinal microbiota. However many questions remain open about their specificities, their mechanism(s) of action and therefore the relevance of their current categorisation.

**Keywords**

Resistant starch • Oligosaccharides • Fermentation • Health effect • Non-digestible carbohydrates • Dietary fibre • Microbiota composition • Competition for substrates • Short-chain fatty acids • Mineral absorption • Appetite regulation • Intestinal barrier • Immune functions

P. Louis (✉) • H.J. Flint
Microbiology Group, Rowett Institute of Nutrition and Health, University of Aberdeen, Foresterhill, Aberdeen AB25 2ZD, UK
e-mail: p.louis@abdn.ac.uk; h.flint@abdn.ac.uk

C. Michel
UMR Physiologie des Adaptations Nutritionnelles, Université de Nantes, INRA, HNB1- CHU-Hotel DIEU, Place Alexis Ricordeau, 44093 NANTES Cedex 1, France
e-mail: catherine.michel@nantes.inra.fr


A. Schwiertz (ed.), *Microbiota of the Human Body*, Advances in Experimental Medicine and Biology 902, DOI 10.1007/978-3-319-31248-4_9

## 9.1 Introduction

The human gastrointestinal tract supports a diverse collection of resident microorganisms (the gut microbiota). Microbial communities in the stomach, small intestine and large intestine all have important impacts on human health, but by far the greatest concentration of microorganisms (predominantly anaerobic bacteria) is found in the large intestine. Microbial growth depends in part on utilization of endogenous host secretions such as mucin, but diet-derived substrates are generally thought to provide the major energy sources for gut microbes. Modification of the diet, especially non-digestible carbohydrates that reach the large intestine, should therefore provide a highly effective approach for modifying the composition and function of the resident gut microbiota. Recent evidence shows that faecal microbiota composition is indeed modified by changing the major non-digestible energy source in the diet (Walker et al. 2011) while the impact of prebiotics on specific groups is well documented (Bouhnik et al. 2004; Flint et al. 2012a).

Molecular analysis of the healthy human faecal microbiota, mainly via 16S rRNA gene sequencing, shows a high degree of diversity and inter-individual variation at the species level. Nevertheless 50–60 species occur at high abundance in most healthy individuals and generally account for more than 50 % of the bacteria present (Tap et al. 2009; Walker et al. 2011; Flint et al. 2012b). Most of these dominant species are representatives of the two most abundant phyla, the Firmicutes and Bacteroidetes, but certain Actinobacteria (*Collinsella aerofaciens* and *Bifidobacterium* spp.) and Verrucomicrobia (*Akkermansia muciniphila*) also show high abundance when data from alternative detection methods such as FISH microscopy (that avoids biases due to PCR and DNA extraction) are considered. The other significant bacterial phylum is the Proteobacteria, which include a number of pathogens. In the past, detailed microbiological work has been focussed on very few of these organisms, with an obvious emphasis on pathogens and on lactic acid bacteria considered to have probiotic potential. In contrast, the microbial ecology of the dominant human colonic anaerobes is in its infancy, but we are beginning to gain an understanding of certain key functions, such butyrate production by Firmicutes (Louis and Flint 2009) and glycan utilization by Bacteroidetes (Martens et al. 2009). Here we will discuss the prebiotic concept in the context of our current knowledge of the gut microbiota and provide a brief overview of the main prebiotics and prebiotic candidates currently being investigated for their effect on the microbiota and associated health effects.

## 9.2 What Is a Prebiotic?

A prebiotic was originally defined by Gibson and Roberfroid (1995) as "a non-digestible food ingredient that beneficially affects the host by selectively stimulating the growth and/or activity of one or a limited number of bacteria in the colon, and thus improves host health". The knowledge of the microbial diversity within the gut as well as the relative abundance of different members of the microbiota was limited at the time, as it was based almost exclusively on culture-based approaches. Bifidobacteria and lactobacilli (designated collectively here as lactic acid bacteria) were generally regarded as the main beneficial components of the microbiota. The best-studied prebiotics, the fructose-based carbohydrates inulin and fructooligosaccharides, showed a strong selective stimulation of bifidobacteria as assessed by culturing, thus a 'prebiotic effect' became synonymous with a 'bifidogenic effect' (e.g., Cummings and Macfarlane 2002). However, the application of emerging molecular tools for microbiota characterisation revealed that culture-based approaches had vastly under-represented some of the more fastidious and most abundant bacteria, and it also provided a better understanding of the level of diversity in the microbiota.

According to the current definition, prebiotics are "selectively fermented ingredients that result in specific changes in the composition and/or activity of the gastrointestinal microbiota, thus conferring benefit(s) upon host health" (Gibson et al. 2010). Thus, prebiotic effects are no longer limited to the colon but may happen anywhere in the gastrointestinal tract.

To date, all known prebiotics are carbohydrates that may also be classed as dietary fibre, so what distinguishes prebiotics from fibre? In order for a compound to be classed as a prebiotic, it has to fulfil three criteria (Gibson et al. 2010):

1. It is resistant to gastric acidity and hydrolysis by mammalian enzymes and gastrointestinal absorption.
2. It can be fermented by intestinal microbiota.
3. It selectively stimulates the growth and/or activity of intestinal bacteria associated with health and wellbeing.

There are several definitions of dietary fibre that are based on different criteria, but a consensus is emerging that carbohydrates with a degree of polymerisation (DP) of $\geq 3$ that are not hydrolyzed by the endogenous enzymes in the small intestine constitute dietary fibre, regardless of their solubility or fermentability (Slavin 2013; Howlett et al. 2010). They include carbohydrates (and lignin) occurring naturally in food as consumed, obtained from food raw material by physical, enzymatic, or chemical means, or synthetic in origin, however, the latter two categories must be shown to have a physiological health benefit (Howlett et al. 2010). Thus, all dietary fibre fulfils the first prebiotic criterion of non-digestibility in the upper gastrointestinal tract, but some do not fulfil one or both of the other two prebiotic criteria. Note that, despite the fact that it is commonly claimed that all prebiotics are dietary fibres (e.g., Slavin 2013), according to the current consensus on dietary fibre there are carbohydrates generally accepted as prebiotics that strictly speaking do not constitute dietary fibre, such as the synthetic disaccharide lactulose (DP $<3$).

In the current prebiotic definition, changes in the gut microbiota are more loosely defined than before, however, the mainstream view still persists that prebiotics are directed at genus level changes and target bifidobacteria and lactobacilli (Gibson et al. 2010). Based on the current knowledge in microbiota composition and activity it has to be questioned however whether this view can be upheld. The following questions in particular need to be addressed.

### 9.2.1 Are Gut Bacteria Either Beneficial or Detrimental for Health?

It is increasingly being recognised that it is too simplistic to categorise bacterial species or groups as either beneficial or detrimental for human health. Thus, a bacterial strain may carry both positive and negative traits and its overall effect on the host may vary depending on the specific gut conditions. For example, *Faecalibacterium prausnitzii* produces butyrate (Duncan et al. 2002), which exerts health-promoting effects on the colonic wall, and also has anti-inflammatory properties that appear to be mediated by factors other than butyrate (Sokol et al. 2008). On the other hand, specific strains of this species also carry high β-glucuronidase activity, which is associated with an increased colorectal cancer risk (McIntosh et al. 2012). The level of generation of carcinogenic compounds originating from gut microbial β-glucuronidase activity is, however, dependent on the level of exposure of the host to the respective precursor molecules, and the level of butyrate produced relative to other fermentation products may vary depending on the gut conditions and activities of other members of the gut microbiota that interact with *F. prausnitzii* in trophic webs. Lactobacilli and bifidobacteria are generally regarded as carrying no traits detrimental to health, however, it is conceivable that an increase in lactic acid bacteria could be detrimental for certain individuals. Some patients with Ulcerative Colitis, for example, have been shown to have very high faecal levels of lactate at the expense of the more health-promoting fermentation acids (Vernia et al. 1988), which may be further exacerbated by the stimulation of lactate producers. Accumulation of D-lactic acid in short bowel syndrome can be life-threatening (Ewaschuk et al. 2005).

### 9.2.2 How Selective Are Prebiotics?

Prebiotics may be less selective than originally thought and directly stimulate non-lactic acid bacteria. It has for example been found that several butyrate producers also have the capacity to degrade certain prebiotics (Scott et al. 2013), and for some of those a significant increase in response to prebiotic intake has been demonstrated in human intervention trials (Ramirez-Farias et al. 2009; Louis et al. 2010; Dewulf et al. 2013). This can easily be missed if microbiota analysis is restricted to a few target groups, or is performed only at a broad phylogenetic scale. Thus, a change in overall abundance of the Lachnospiraceae within Firmicutes may not be found in response to prebiotic intake, while specific sub-groups or species within this diverse family may show a response. We therefore require further studies that determine microbiota changes at a fine scale (see eg. Chung et al. 2016). Based on the current evidence it seems likely that few if any prebiotics will be entirely specific to bifidobacteria or lactobacilli, or any other bacterial species or genus. This is further complicated by the fact that different individuals may show different responses to the same prebiotic (Figs. 9.1 and 9.2), which is often masked by the fact that many studies only report mean data.

### 9.2.3 Should Prebiotic Effects Be Limited to Specific Species or Genera Within the Microbiota?

Gut bacteria exist within a complex community with extensive interactions between the individual members, thus their actions cannot be seen in isolation. Higher butyrate production has been reported in response to prebiotic consumption (Gibson et al. 2010), but neither lactobacilli nor bifidobacteria produce this fermentation acid. They may, however, indirectly increase butyrate production by feeding lactate to lactate-utilising butyrate producers (Duncan et al. 2004). It is also being discussed that if gas production accompanies prebiotic intake, the carbohydrate is not acting as an authentic prebiotic, as bifidobacteria and lactobacilli cannot produce gas (Gibson et al. 2010), but again, these bacteria do not exist in isolation and their activities may stimulate other gas-producing bacteria. Any prebiotic is therefore likely to lead to effects that are an indirect consequence of stimulation of lactic acid bacteria and it appears arbitrary not to reject a prebiotic that leads to positive associated effects (such as butyrate production) but reject one that leads to negative ones (such as gas production). In fact, gas production is commonly found to co-occur for carbohydrates generally accepted as prebiotics and it appears unfeasible to define a prebiotic based on the intake levels that don't lead to such effects, especially as this differs widely between different individuals.

### 9.2.4 Might the Beneficial Effects of a Prebiotic Be Due to Microbiota Changes Other than Stimulation of Lactic Acid Bacteria?

If prebiotics are found to stimulate more than one group of bacteria, then it becomes difficult to assign their health benefits to a single group of 'beneficial' bacteria. This dilemma is highlighted for example by work on the effects of inulin. In the study of Ramirez-Farias et al. (2009), dietary inulin was found to stimulate both bifidobacteria and *F. prausnitzii* in stool samples from healthy subjects. Since *F. prausnitzii* is a butyrate-producer with claimed anti-inflammatory action (Sokol et al. 2008) it could easily be argued that the increase in its population, rather than that of bifidobacteria, would explain any health benefit. Indeed, many currently obscure, but numerically significant, anaerobes could become candidates for delivering health benefits once we know more about their biology and their interactions with the host.

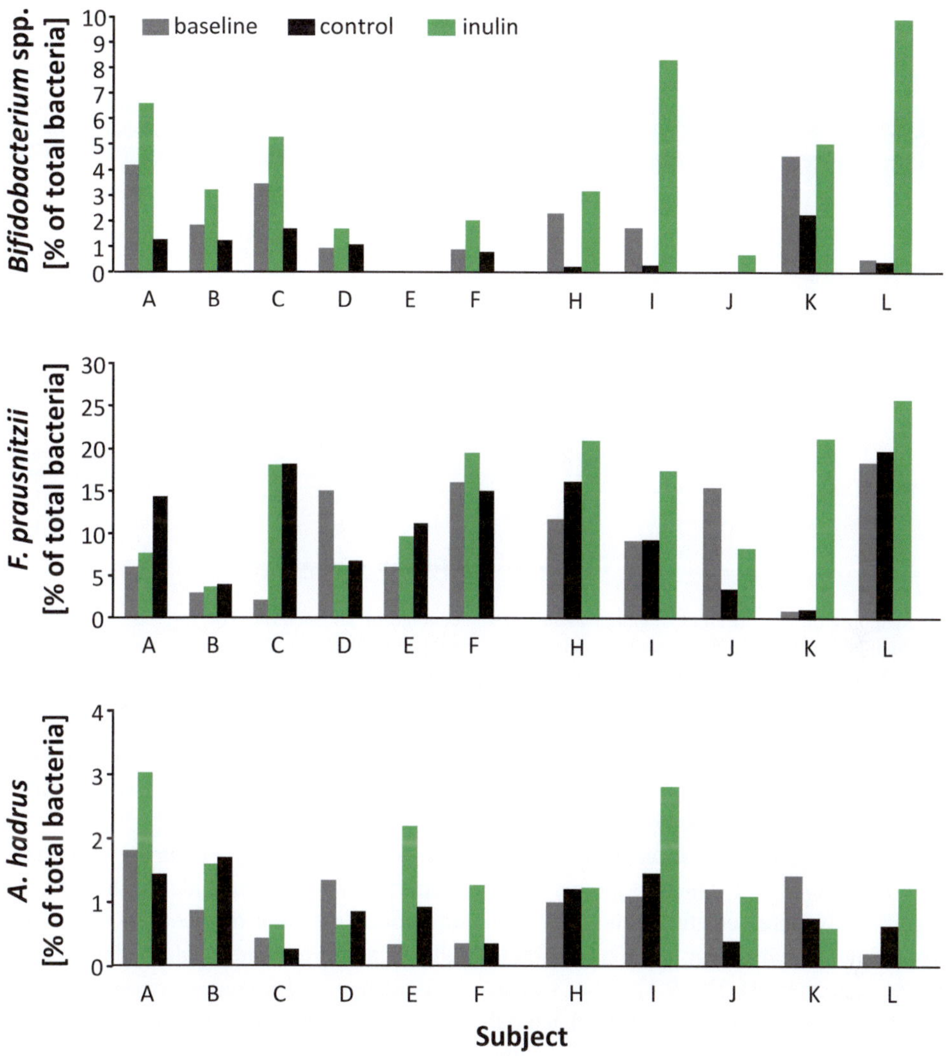


**Fig. 9.1** Three bacterial groups showing a significant mean increase in response to inulin intake in a human volunteer trial, determined by quantitative PCR against the 16S rRNA gene as a percentage of all bacteria. The response is shown for each subject individually. Level of significance across all volunteers: *Bifidobacterium* spp. $P<0.001$, *F. prausnitzii* $P=0.019$, *A. hadrus* $P=0.003$ (Ramirez-Farias et al. 2009; Louis et al. 2010)

### 9.2.5 Does Variation in Microbiota Responses Between Individuals Need to Be Considered?

There is substantial inter-individual variation in the species composition of the gut microbiota. Walker et al. (2011) found significant responses to controlled dietary manipulation for a subset of 'diet-responsive' bacterial species. On the other hand, overall microbiota composition was driven more by the individual than by the diet (Walker et al. 2011; Salonen et al. 2014). Furthermore, individuals showed variation in the response of their microbiota to the dietary change that appeared to be related to the species composition of the microbiota at the start of the intervention period (Walker et al. 2011). There is evidence that such variation applies equally to prebiotic interventions. Ramirez-Farias et al. (2009) found that,

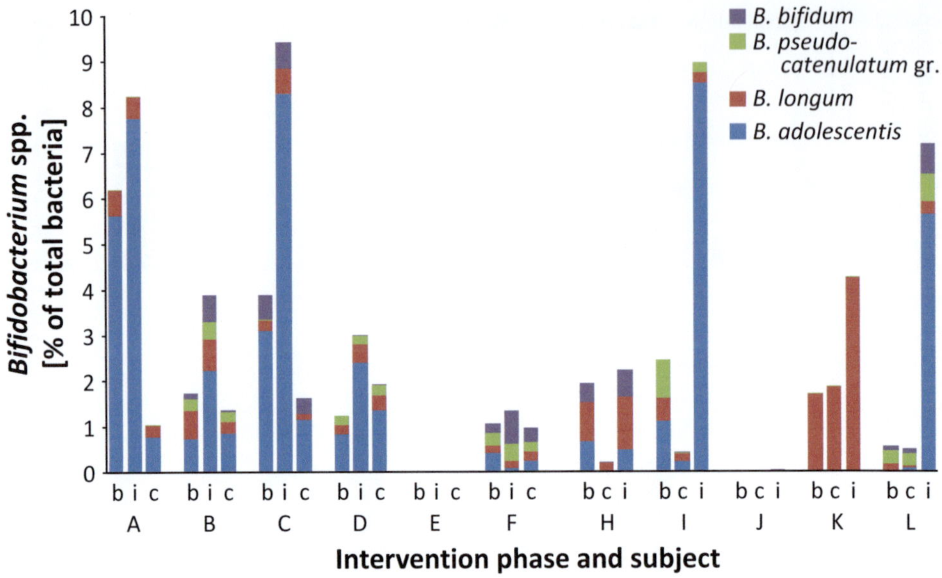


**Fig. 9.2** Response of four different *Bifidobacterium* species to inulin intake in a human volunteer trial, determined by quantitative PCR against the 16S rRNA gene as a percentage of all bacteria. The response is shown for each subject individually. Intervention phase *b* baseline, *c* control period, *i* inulin period (Ramirez-Farias et al. 2009)

despite a significant average increase in bifidobacteria following inulin supplementation, certain individuals who showed very low initial bifidobacterial numbers did not exhibit a bifidogenic response (Fig. 9.1). Two Firmicutes groups found to be significantly increased after inulin consumption also did not reveal a clear trend in relation to initial relative abundance (Fig. 9.1). Other studies however have reported a greater bifidobacterial response when the initial population is low (Gibson et al. 2010). Ramirez-Farias et al. (2009) found that *Bifidobacterium adolescentis* showed the greatest mean response, but some individuals showed an increase in other *Bifidobacterium* species (Fig. 9.2). If these bacterial changes have consequences for health (as is implied by the prebiotic definition) then we must expect there to be inter-individual variation in health benefits. It should also be noted in this context that prebiotics do not always elicit the same effects in patients as they do in healthy people (Whelan 2013). It may be that such variation would be less extreme when effects are mediated through common metabolic products than when they depend on interactions of cell components with the immune system, which may even be strain-specific.

### 9.2.6 Is Microbiota Diversity Itself an Indicator for Gut Health?

The overall microbiota diversity may also be an important factor in considering health effects of prebiotics. A more diverse ecosystem is generally regarded as more resilient due to functional redundancy of different members of the community, and reduced diversity has been observed in certain disease states (Lozupone et al. 2012). Two studies determining bacterial richness by quantitative metagenomics recently found that human populations broadly fall within two categories of low or high faecal gene count, with volunteers in the reduced gene richness category showing higher levels of anthropometric and biochemical phenotypes associated with disease, as well as microbial metabolic pathways more associated with an inflammation-associated status. Intriguingly, the two states appeared to correlate with abundance changes in relatively few bacterial species (Cotillard et al. 2013; Le Chatelier et al. 2013). More work is clearly necessary to fully understand the implications of microbiota diversity and its effect on human health, but it has recently been suggested to include the stimulation of ecological biodiversity in the definition of prebiotics (Van den Abbeele et al. 2013). If high diversity is indeed a health-promoting factor, high-level intake of a specific prebiotic may actually be more detrimental to health than consuming fibre from a variety of sources.

The above considerations make it clear that the complexity of the microbiota and its interactions with the host, as well as differences between individuals, make it difficult to define what constitutes a healthy gut microbiota. This has recently been pointed out by the European Food Safety Authority (EFSA). Their guidance document on scientific requirements for health claims related to gut function states that "based on current scientific knowledge, it is not possible to define the exact numbers of the different microbial groups which constitute a normal microbiota" and that the evidence available "does not establish that increasing the number of any groups of microorganisms, including lactobacilli and/or bifidobacteria, is in itself a beneficial physiological effect" (EFSA panel on dietetic products, nutrition and allergies 2011). Furthermore, recent developments illustrate how much the definition of prebiotics is closely dependent on the evolution of knowledge about the nature and the possible benefit of interactions between bacterial species that constitute the intestinal microbiota and the host. Accordingly, it is highly probable that this definition will change further in the future.

## 9.3 Mechanisms of Action of Microbiota Modulation by Prebiotics (Fig. 9.3)

The most obvious mechanism by which a prebiotic can alter gut microbiota composition is by providing an energy source that can only be accessed by certain members of the microbial community. This would be expected to lead to enhanced growth of selected microorganisms and to their increased representation within the community (Flint et al. 2007). Growth tests using cultured strains show that phylogenetically distant bacterial species often share the ability to utilize a given prebiotic (Scott et al. 2013). This was also demonstrated recently by a functional metagenomics approach that identified genes from a human microbiota metagenomic library

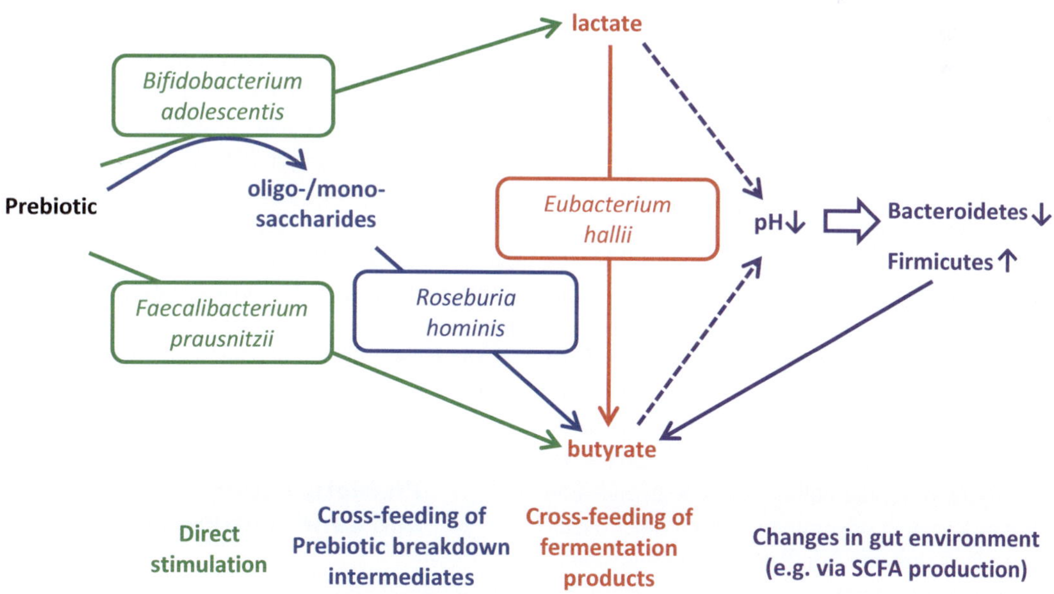


**Fig. 9.3** Possible mechanisms of prebiotic action on the gut microbiota, illustrated with the example of butyrate generation in response to prebiotic intake. For further detail see main text

for breakdown of several prebiotics in the heterologous host *Escherichia coli* (Cecchini et al. 2013). Clones originating from several different Firmicutes, Actinobacteria and Bacteroidetes were able to degrade fructooligosaccharides, xylooligosaccharides, galactooligosaccharides or lactulose. At the same time, utilization can be species-specific within a given genus, and strain-specific within a given species. This is clearly seen for *Bifidobacterium* spp. in respect of utilization of starch (Belenguer et al. 2006; Ryan et al. 2006) and fructans (Rossi et al. 2005). Chain length is an important factor in determining selectivity, e.g., among fructans very few species are able to utilize long chain inulin, whereas utilisation of short chain fructooligosaccharides is widespread (Scott et al. 2013). Growth in pure culture however does not equate to competitive success within the complex gut community following addition of a prebiotic. Carefully monitored human dietary trials are therefore crucial in defining prebiotic outcomes (Ramirez-Farias et al. 2009). In time, genome sequence information should help to explain and predict inter-species and inter-strain differences in carbohydrate utilization, but current understanding of the mechanistic basis for substrate utilization in most human colonic anaerobes is surprisingly limited (Flint et al. 2012b).

The selective stimulation of certain microbial populations by a prebiotic means that their metabolic products are likely to be produced at an increased rate. Indeed it is theoretically possible for a prebiotic to cause increased product formation without an increase in the relevant bacterial population, through simple mass action. There is therefore considerable potential for indirect stimulation of other organisms within the microbial community that may be able to benefit from these metabolites. Lactate is a major product of fermentation for many gut bacteria, including the lactic acid bacteria, lactobacilli and bifidobacteria, when grown in pure culture. Lactate can also be used as a growth substrate by certain other bacterial species that are abundant in the human colon, to produce butyrate (Duncan et al. 2004; Morrison et al. 2006) or propionate (Bourriaud et al. 2005). The relatively low concentrations of lactate that are detected in healthy human faecal samples may be explained by the efficient consumption of lactate (Belenguer et al. 2006) with lactate accumulating only when the activity of the lactate-utilizers is compromised, for example by reduced pH (Belenguer et al. 2007). Thus a stimulation of lactate-producing species might lead to increased populations of lactate-utilizing species such as *Eubacterium hallii* (Belenguer et al. 2007), and to increased production of butyrate or propionate. Another form of cross-feeding involves the release of partial breakdown products from complex substrates by one species, making them available for utilization by other species (Belenguer et al. 2006; Falony et al. 2006). For example it has been demonstrated recently that several other species can benefit from the ability of *Ruminococcus bromii* to degrade particulate resistant starches (Ze et al. 2012). Stimulation of specific members of the microbiota may also have antagonistic effects on others, such as enhanced production of bacteriocins.

Prebiotics may also alter the gut environment. Most obviously, their fermentation will tend to decrease the pH of the gut lumen as a result of increased production of fermentation acids. In vitro work with pH-controlled, continuous flow fermentors has shown that a one unit shift in pH between 5.5 and 6.5 can greatly alter the species composition of the gut microbiota and consequently its metabolic products (Walker et al. 2005; Duncan et al. 2009). Thus another possible explanation for the butyrogenic effect of a prebiotic is that a decrease in gut pH promotes particular butyrate-producing Firmicutes by decreasing competition with the more acid-sensitive *Bacteroides* spp. (Walker et al. 2005). In addition to pH, prebiotics might also affect the gut environment through other effects, eg. viscosity, gut transit and interactions with other food components.

## 9.4 Ingredients with Confirmed Prebiotic Action and Candidate Prebiotics

There are several comprehensive reviews on prebiotic effects of various carbohydrates, especially fructans and galactooligosaccharides, which are generally regarded as carbohydrates with confirmed prebiotic properties (for example, Macfarlane et al. 2008; Gibson et al. 2010; Roberfroid et al. 2010; Whelan 2013). Here we will provide a brief overview of the structure of the main classes of (candidate-) prebiotics currently being investigated as well as their natural occurrence in food ingredients and/or commercial synthesis, and review some recent results on their effects on the gut microbiota.

### 9.4.1 Fructans

Inulin and fructooligosaccharide (FOS, alternatively named oligofructose) consist of linear fructose chains in $\beta(2\rightarrow1)$ linkage, usually with a terminal glucose unit in $\beta(2\leftrightarrow1)$ linkage as in sucrose (Fig. 9.4). Inulin is normally present as a variety of chain lengths with a DP of up to 60, whereas FOS has a DP of less than 10. FOS can be manufactured from inulin by partial hydrolysis or synthesized enzymatically (Gibson et al. 2010). Inulin-type fructans are present in relatively high quantities in various vegetables (e.g., chicory root, Jerusalem artichoke), but they are also found in smaller amounts in cereals such as wheat. Due to high intake levels of cereal products (e.g., bread), they are the major contributor to fructan intake in Western societies (Whelan 2013). While originally thought to specifically stimulate lactic acid bacteria, there is now evidence from several independent studies that other bacterial species may also be stimulated either directly or indirectly, as detailed in the sections above. Chain length also appears critical for determining which bacteria can act as primary degraders of fructans (Scott et al. 2013).

### 9.4.2 Galactooligosaccharides

Pulses are rich in natural galactooligosaccharides (GOS), including raffinose family oligosaccharides (RFO), which are based on the extension of sucrose with galactose residues (Fig. 9.4) (Johnson et al. 2013; Whelan 2013). Raffinose and stachyose are also known as soybean oligosaccharides, but their impact on the gut microbiota has not been established in sufficient detail. GOS based on lactose, and containing an extra galactose residue at $C_3$, $C_4$ or $C_6$, are present in human milk. GOS is also produced by enzymatic transglycosylation of lactose, which results in mixtures of mostly tri- to pentasaccharides with galactose in $\beta(1\rightarrow6)$, $\beta(1\rightarrow3)$ and $\beta(1\rightarrow4)$ linkage (Fig. 9.4), with the exact composition depending on the enzymes used and reaction conditions. These transgalactooligosaccharides are also known as TOS (Macfarlane et al. 2008; Gibson et al. 2010). Stable isotopically labeled GOS was used recently to examine the selectivity of this prebiotic in an in vitro model of the proximal colon (Maathuis et al. 2012). Several bifidobacteria and lactobacilli showed the highest label incorporation, with infant-type bifidobacteria seemingly stimulated most strongly. Other bacteria, including members of the Enterobacteria, Bacteroidetes and Firmicutes also incorporated the label to a lesser degree.

Lactulose is a synthetic disaccharide produced by the isomerisation of lactose (Fig. 9.4). With a DP of two it does not fulfil the criteria as a dietary fibre, but it is generally regarded as a prebiotic (Gibson et al. 2010). Lactulose-derived GOS is more resistant to upper gut digestion than lactose-derived GOS. A comparison of both types of GOS in a rat model revealed a significant stimulation of the Firmicutes *Eubacterium rectale/ Clostridium coccoides* group in caecum and colon, whereas bifidobacterial stimulation reached significance only for lactulose-derived GOS in both intestinal compartments, with lactobacilli being significantly increased on lactose-derived GOS in the colon only (Marín-Manzano et al. 2013). The differences seen between the two types of GOS may be due to their difference in indigestibility and/or structural differences (such as types of glycosidic linkages).

### 9.4.3 Resistant Starch, Starch- and Glucose-Derived Oligosaccharides

Resistant starch (RS) is the fraction of starch (Fig. 9.4) that resists digestion in the upper gut

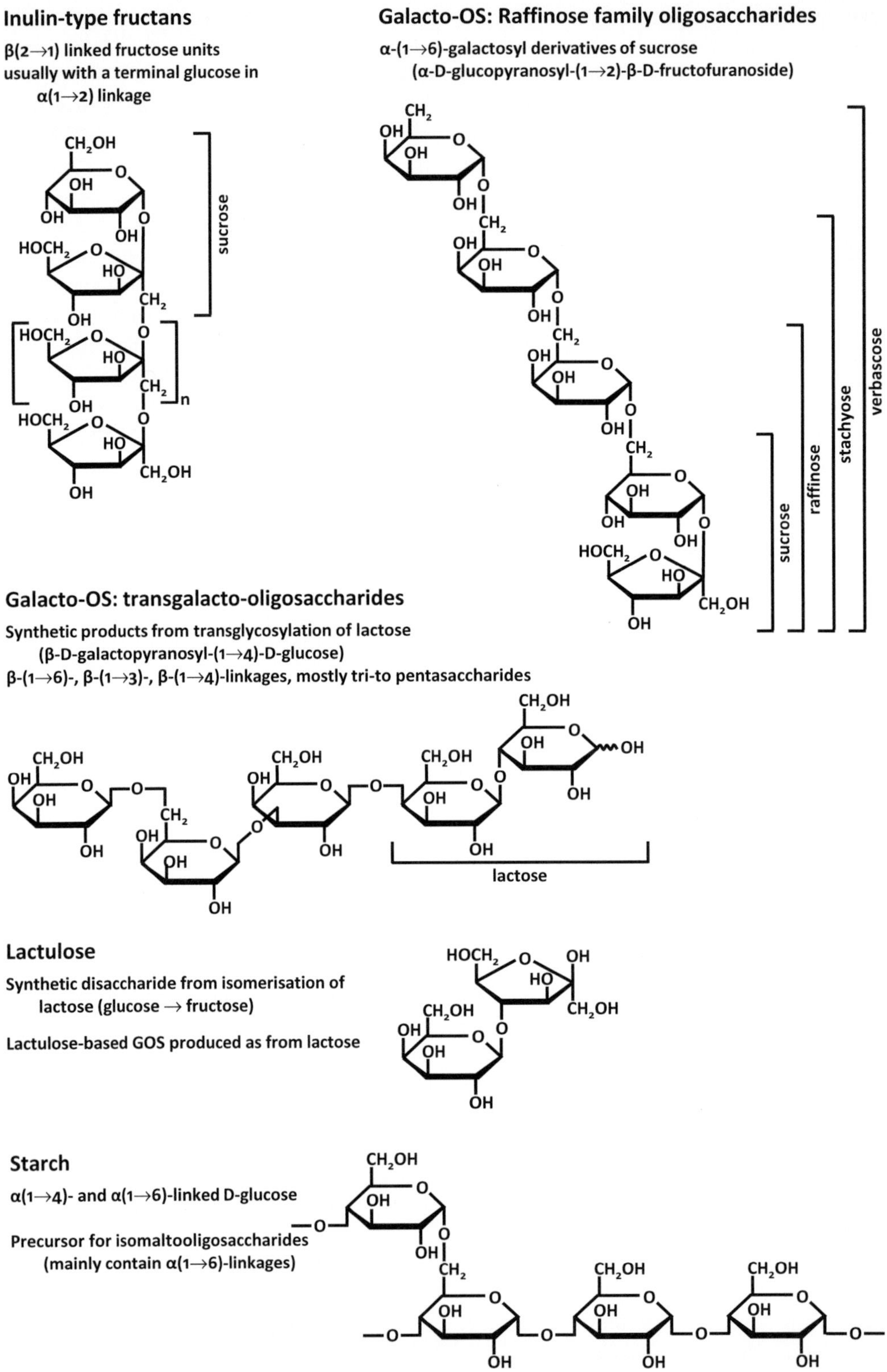


**Fig. 9.4** Carbohydrate structures of fructans, galactooligosaccharides and starch

due to its physicochemical properties. The amount of RS varies between different foods (such as legumes, cereals and potatoes), and food processing can also affect RS levels. While not currently regarded as a prebiotic universally (e.g., Gibson et al. 2010), its classification as a prebiotic has been proposed, particularly as RS intake is associated with health-promoting effects, such as high butyrate formation (Fuentes-Zaragoza et al. 2011). Selective stimulation of several bacterial groups belonging to the Firmicutes in response to a diet high in RS has recently been demonstrated (Walker et al. 2011). Pure culture work in vitro confirmed the starch-degrading ability of *R. bromii* and *B. adolescentis*, and to a lesser degree of *E. rectale* and *Bacteroides thetaiotaomicron*. However, in mixed bacterial and faecal incubations, the presence of *R. bromii* was crucial for efficient RS breakdown (Ze et al. 2012). This was in agreement with the observation that human volunteers with very low levels of *R. bromii* were not able to digest dietary resistant starch completely (Walker et al. 2011).

Isomaltooligosaccharides (IMO) are found naturally in some fermented foods and honey, and they are commercially produced from starch by enzymatic hydrolysis, resulting in glucans containing mainly $\alpha(1\rightarrow6)$ linkages, although other linkages are also present, and other types of glucooligosaccharies (GOS) are also considered as IMOs (Goffin et al. 2011). IMOs are widely used in the Asian market. They are partially digestible, with digestibility depending on DP and type of linkages present (Goffin et al. 2011). There is some indication that IMOs selectively stimulate lactic acid bacteria (Gibson et al. 2010; Goffin et al. 2011), however, studies investigating changes in the whole microbiota are required to fully evaluate their effect on the microbiota.

Polydextrose, a highly branched glucan containing a range of glycosidic linkages, is synthesized from glucose. There is some evidence that it is selectively fermented by bifidobacteria, however, a human intervention study that looked at a wider range of bacterial groups using molecular methods did not confirm this (Costabile et al. 2012) and further work is required to establish its effects.

### 9.4.4 Other Oligosaccharides

Pectic oligosaccharides (POS) are derived from pectin, a polysaccharide present in various fruits and vegetables (e.g., citrus fruit, apple, sugar beet). The polysaccharide backbone consists of galacturonic acid residues (homogalacturonan) that may alternate with rhamnose residues (rhamnogalacturonan I). The carboxyl groups may be modified by methyl esterification, and $C_2$ or $C_3$ positions may be acetylated. Sidechains containing various sugars (arabinose, galactose, xylose etc.) are often present and may be substituted with ferulic acid (Yoo et al. 2012). In vitro and in vivo studies to establish a prebiotic effect lead to mixed results, which may be complicated by the fact that POS from different sources show structural variation (Gullón et al. 2013). Known pectin degraders include several *Bacteroides* spp. as well as the Firmicutes *Eubacterium eligens* and *F. prausnitzii* (Lopez-Siles et al. 2012).

The production and characterisation of novel oligosaccharides from non-digestible carbohydrate fractions of biomass, such as xylo-, arabino- and mannooligosaccharides is an active research area and complex foodstuffs are also being investigated as sources of novel prebiotics (e.g., Otieno and Ahring 2012; Yoo et al. 2012). The increasing use of genomic and metagenomic data should enable a more rational design of prebiotic compounds in the future (Candela et al. 2010).

### 9.4.5 Non-carbohydrate Compounds

The definition of prebiotics is not restricted to carbohydrates per se (Gibson et al. 2010), however, carbohydrates appear most likely to fulfil the criteria of non-digestibility and selective fermentability. Cocoa-derived flavanols have also been proposed as prebiotics, as they elicited a stimulatory effect on lactic acid bacteria in vivo and in vitro (Tzounis et al. 2011). It is unlikely, however, that this effect is due to fermentation of those compounds, in the sense that the corresponding bacteria achieve a substantial energy gain, for the following reasons: intake levels are generally much lower than for carbohydrate-based prebiotics and a proportion of the phenolics will be absorbed by the host either directly or after biotransformation by the microbiota. Furthermore, the underlying biochemistry of the bacterial metabolic transformations (cleavage of complex to more simple phenolics, hydrogenation, demethylation, dehydroxylation, etc., Russell and Duthie 2011) is unlikely to result in a major energy gain for the microbes performing the transformation. Phenolic compounds may, however, exert antimicrobial effects, the potency of which may differ between bacterial species (Louis and O'Byrne 2010), which can lead to selective stimulation of certain bacterial groups. The microbial transformation of phenolics may indeed function as a detoxification mechanism for the bacteria. This raises the question whether the second prebiotic criterion, fermentability by the intestinal microbiota, should be retained, or whether it ultimately does not matter by which mechanism selective changes within the microbiota are elicited.

## 9.5 Health Effects of Prebiotics on the Host

By definition, prebiotics are supposed to exert beneficial impacts on host health and/or well-being. This criterion has motivated many studies on the ability of prebiotics to induce physiological effects, many of which are now proven, at least in the case of inulin and/or FOS, on which the majority of these studies had focused (see summary in Table 9.1 and, for extensive review: British Journal of Nutrition Vol. 93, Suppl. 1 2005; Journal of Nutrition Vol. 137 Suppl. 2007; Roberfroid et al. 2010). The transferability of these findings to other candidate prebiotics is being evaluated.

The definition of prebiotics also assumes that the health benefits result from their impact on the composition and/or activity of the gastrointestinal microbiota, i.e., their fermentation. Prebiotic fermentation in the large bowel is expected to result in changes in the whole ecosystem, namely increase in the total bacterial mass, stimulation/inhibition of some particular bacterial strains, production of numerous bacterial metabolites, among which organic acids predominate and – as a consequence of the latter – acidification of the luminal contents. All these features can exert biological effects (Macfarlane and Macfarlane 2012; Russell et al. 2013), which, according to their nature and combination, can result in different physiological effects. However, for most of the physiological effects induced by prebiotics, the exact contribution of each of these events and the precise nature of bacterial factors and of mechanistic pathways involved are still to be deciphered.

Finally, whether these physiological effects translate into actual improvements in terms of prevention and/or curing of diseases related to the consumption of prebiotics is still a matter of debate. When assessing health effects of prebiotics it also has to be kept in mind that a correlation between gut microbial changes and health markers may be by mere association rather than by a causal relationship. Some prebiotics may exert their health-promoting effects independent of a promotion of beneficial bacteria, for example by direct stimulation of the immune system via host receptors, or by pathogen binding, which reduces pathogen adherence (Licht et al. 2012).

### 9.5.1 Physiological Effects and Underlying Mechanisms

#### 9.5.1.1 Improvement of Intestinal Functions (Stool Bulking, Stool Regularity, Stool Consistency)

In babies and possibly (but with contradictory results in this case) in infants, supplementation with a GOS/inulin mix increases stool frequency, soften stools, acidifies stool pH, and modulates the SCFA pattern similar to that of breast-fed infants (Gibson et al. 2010; Roberfroid et al. 2010; Tabbers et al. 2011). This was also observed with polydextrose, which is a prebiotic candidate (Ashley et al. 2012). However, in adults inulin has little effect on stool weight (Slavin 2013). FOS can cause symptoms, including bloating,

**Table 9.1** Hypothetical mechanisms underlying the biological and physiological impacts of fructans

| Biological impact (evidence level) | Physiological impact | Expected impact on health | Suggested mechanism: Linked to modification of the microbiota composition | Suggested mechanism: Linked to stimulation of SCFA production | Suggested mechanism: Other |
|---|---|---|---|---|---|
| Increased calcium absorption (+++) | Increased bone density (+, particularly in teenagers) | Osteoporosis prevention | Stimulation of colonocyte $Ca^{2+}$ capture by some bacteria | Trophicity | Increased calcium solubility due to fermentation-induced luminal acidification |
| | | | | Increased mucosal expression of calbindin | |
| Increased production of satietogenic gut peptides (GLP-1) (++) | Appetite regulation (+, few studies) | Obesity treatment | | Stimulation of the mucosal expression of proglucagon (→GLP-1) | |
| Decreased plasma triglycerides and cholesterol (+ in hyperlipidemic humans; +/– in healthy subjects) | Lipidemia regulation (+ in hyperlipidemic humans; +/– in healthy subjects) | Cardiovascular disease prevention | Chelation and/or bile salt hydrolysis by some bacteria | Through impact on GLP-1? | Decreased reuptake of bile salts due to acidification of colonic contents |
| | | | | Suppression of hepatic lipogenesis by propionate | |
| Trophicity of the colonic mucosa | Intestinal barrier homeostasis | Inflammatory bowel diseases (+ colorectal cancer, obesity and diabetes?) treatment | Stimulation of mucus production by some bacteria | Main energy source for colonocytes | |
| Increased mucus production | | | | Stimulation of proliferation and cell differentiation | |
| Decreased intestinal permeability (+) | | | | Regulation of tight-junctions | |
| | | | | Stimulation of the mucosal expression of proglucagon (→GLP-2) | |
| Increased IgA and IL-10 production, phagocytosis and NK activity | Immunomodulation | Miscellaneous: Treatment of infections Allergy prevention, IBS/IBD prevention and treatment colorectal cancer prevention | Immunomodulatory properties of some bacteria | Immunomodulatory properties of butyrate | Direct interaction between fructans and immune cells |
| Decreased inflammatory response (++) | | | | | |

flatulence, and soft stools in adults when fed at high doses (>10 g/day) (Brownawell et al. 2012; Slavin 2013).

From knowledge accumulated on dietary fibre, improvement of stool bulking is the result of increased stool weight due to either the physical presence of the fibre and to the water held by the fibre, or an increased bacterial mass, with the former phenomenon being much more effective than the latter one (Brownawell et al. 2012; Slavin 2013). In this context, one has to assume that the different stool bulking effects between infants and adults result from differences in fermentation intensity (related to microbiota immaturity in infants): the stool bulking capability of prebiotics in infants thus appear to stem from the fact that they may increase the stool water content by osmolarity, because they are likely to be incompletely fermented. In agreement with this assumption, residual unfermented prebiotics have been detected in infants stools (Moro et al. 2005), whereas prebiotics are totally degraded in adults.

#### 9.5.1.2 Stimulation of Mineral Absorption and Improvement of Bone Density

Numerous studies carried out in animals have demonstrated that linear fructans, particularly a mixture of inulin and FOS, improve mineral (and especially calcium) absorption. Some have also shown that this results in increases in whole body bone mineral content and in bone density mass (Scholz-Ahrens et al. 2007; Roberfroid et al. 2010).

In humans, these conclusions must be qualified according to the age of the subjects, their hormonal status, their calcium intake and of course the prebiotic dosing. Indeed, the few studies investigating the impact of baby formulae supplemented with prebiotics (mixture of GOS/inulin, 9/1) on calcium absorption or other markers of bone mineral metabolism did not reveal any change (Yap et al. 2005; Hicks et al. 2012). By contrast, in teenagers, a mix of inulin and FOS not only enhanced calcium absorption, but also calcium accretion in bones (Abrams et al. 2005; Roberfroid et al. 2010). Contrasting results have been obtained in adults. In postmenopausal women it appears to depend on the number of years past the onset of menopause, with no effect in early postmenopausal women, but calcium absorption improvement in women who are in the late postmenopausal phase (Roberfroid et al. 2010). The beneficial impact of fructans on calcium absorption is evidenced only if the calcium intake is sufficient (Scholz-Ahrens and Schrezenmeir 2002) and appears to depend on the fractional calcium absorption at baseline, with those individuals with lower absorption before treatment showing the greatest benefit (Griffin et al. 2002). Finally, the prebiotic dosage also influences the benefit and according to Roberfroid et al. (2010), a minimum level of 8 g/day seems to be required to elicit an improvement on both calcium absorption and bone mineralisation.

Whether or not such effects are expected for all prebiotics is not clear. On the one hand, GOS appears promising for improving calcium absorption in both animal and human studies (van den Heuvel et al. 2000; Weaver et al. 2011) and various (candidate-) prebiotics such as soyaoligosaccharides, lactulose, or resistant starch have also provided evidence of a positive effect on calcium absorption, at least in the rat (Roberfroid et al. 2010). On the other hand, fructans with different degrees of polymerisation (average DP=3–4 *vs* average DP>23 *vs* mix of the two), have different effects on calcium retention, femoral bone density and bone calcium content in rats (Griffin et al. 2002; Kruger et al. 2003).

Similarly, generalization of these benefits to other minerals is not possible, even if the impact of prebiotic consumption on the metabolism of phosphorous, magnesium, iron, copper, and zinc has occasionally been considered. A beneficial impact on magnesium absorption seems likely (Yap et al. 2005; Scholz-Ahrens and Schrezenmeir 2002; Roberfroid et al. 2010; Legette et al. 2012), even if available data are very limited. Iron as well as zinc absorption may also be improved, while retention of phosphorous appears not to be affected (Scholz-Ahrens and Schrezenmeir 2002).

The exact mechanisms involved in the improvement of calcium absorption are not fully deciphered. They may involve acidification of the lumen content which increases calcium solubility (Scholz-Ahrens and Schrezenmeir 2002). Short-chain fatty acids (SCFA) also have a trophic effect on the mucosa (Hamer et al. 2008), which may lead to enlargement of the absorption surface. Enhanced butyrate or propionate production also stimulates calbindin-D9k expression by colonocytes (Fukushima et al. 2012), a feature which is observed in rats consuming FOS (Ohta et al. 1998). Other hypotheses are related to stimulation of particular bacterial species which may improve either calcium bioavailability (Bergillos-Meca et al. 2013) or calcium absorption by colonocytes (Gilman and Cashman 2006), or which may stimulate the colonic production of equol, a phytooestrogen which is associated with bone health (Coxam 2007).

#### 9.5.1.3 Regulation of Appetite and Stimulation of Gut Peptide Secretion

Numerous studies have shown that fructan consumption decreases energy intake in rodents (Roberfroid et al. 2010; Delzenne et al. 2013), although this was only seen in male animals but not females at a specific time point in a long-term study investigating the effect of lifelong intervention (Rozan et al. 2008). Fructan intake is usually accompanied by an increase in glucacon-like peptide 1 (GLP-1) and, although less documented, by a increase in peptide YY (PYY), two anorexigenic peptides secreted by the intestine (Roberfroid et al. 2010; Delzenne et al. 2013). The crucial role of GLP-1 has been demonstrated using GLP-1 receptor knockout mice (GLP-1R(−/−)) in which the FOS-induced decrease of energy intake was abolished (Cani et al. 2006a).

In humans, pre-adaptation to FOS induced changes in satiety and reduced total energy intake per day (Cani et al. 2006b; Parnell and Reimer 2009) but this does not appear to hold up for acute fructan supplementation (Peters et al. 2009; Hess et al. 2011). Anyhow, when occurring, decreased energy intake in humans was also related to an increase in satietogenic and/or a decrease in orexigenic (ghrelin) peptides (Delzenne al. 2013).

The capability of other prebiotics to reduce energy intake and modulate gut peptides is poorly described, however, a recent study suggests that this property is shared by GOS and by oligosaccharides derived from arabinoxylan (AXOS): rats consuming GOS presented reduced energy intake and increased gene expression of PYY and proglucagon, the precursor for GLP-1 (Overduin et al. 2013) and similar results were obtained in mice fed AXOS (Neyrinck et al. 2012).

The underlying mechanism for these effects is supposed to be a stimulation of L-endocrine cells in the intestine, either through triggering differentiation of these cells or through stimulated expression of gut peptides by these cells (Delzenne et al. 2013). SCFA and particularly butyrate possibly mediate such a stimulation, since this bacterial metabolite seems most effective in stimulating GLP-1 production in vitro (Zhou et al. 2008). The free fatty acid receptors FFAR2 (GPR43) and FFAR3 (GPR41), which recognize SCFA, may be involved in the stimulation of GLP-1 secretion (Tolhurst et al. 2012). Finally, a new finding suggests that FOS consumption results in changes in the neuronal activation of the arcuate nucleus, an hypothalamic structure which contributes to the control of food intake (Anastasovska et al. 2012) but this observation could result from gut peptide stimulation, since functional receptors for GLP-1 are also expressed by hypothalamic neurons (Dalvi et al. 2012). It should be noted that different types of dietary fibre not currently classed as prebiotics also regulate appetite, which may partially be due to physicochemical effects in the gastrointestinal tract (Slavin and Green 2007).

#### 9.5.1.4 Improvement of Intestinal Barrier Integrity

In adult animals, fructan fermentation has been shown to affect the mucus layer and the intestinal mucosal morphometry by increasing the height of villi and the depth of the crypts (Kleessen and Blaut 2005). FOS supplementation decreases intestinal permeability and improves tight-junction integrity in mice (Cani et al. 2009) and this improvement in intestinal permeability has been confirmed for humans (Russo et al. 2012). These effects appear to be specifically induced by fructans since the few studies dedicated to other prebiotics (GOS or mix of different OS) failed to demonstrate any particular effect on the intestinal barrier (Meslin et al. 1993; Barrat et al. 2008; Westerbeek et al. 2011).

Low-grade systemic inflammation (also called metabolic endotoxemia) with elevated serum levels of bacterial lipopolysaccharides (LPS) is often observed in obese patients, especially when they also present with metabolic disorders. An improvement of the intestinal barrier is likely to contribute to the reduction in metabolic endotoxemia that has been reported in response to prebiotic intake in animal models (Delzenne et al. 2013).

Numerous biological mechanisms may participate in the beneficial impact of fructans on intestinal barrier. First, specific fructan-stimulated bacteria may be involved since numerous bacterial strains used as probiotics have been shown in vitro to increase mucin expression and to enhance tight junction stability, which decreases epithelial permeability (Ohland and Macnaughton 2010). Second, butyrate may also contribute since it improves mucosal trophicity and it can also regulate mucin production and cellular permeability in vitro and ex vivo (Hamer et al. 2008). Third, glucagon-like peptide 2 (GLP-2), which is co-secreted with GLP-1 by endocrine L cells, appears to be a key factor in the regulation of the intestinal barrier, since the effects of FOS on intestinal barrier were reproduced by pharmacological GLP-2 treatment and abolished in the presence of a GLP-2 antagonist (Cani et al. 2009, 2012). Furthermore, prebiotics may also exert an influence on the endocannaboid system with consequent effects on gut barrier function (Delzenne et al. 2013). Again these mechanisms can be interconnected, considering that the stimulatory effect of butyrate on GLP-1 secretion mentioned above is likely to occur for GLP-2 as well, since both derive from the same proglucagon gene.

#### 9.5.1.5 Regulation of Lipid and Glucose Metabolism

Fructans improve glucose homaeostasis in several rodents (see for review Roberfroid et al. 2010). With regard to lipid metabolism, fructans have been shown to decrease total plasma cholesterol in mice or rats and to decrease triglyceride plasma concentration (and, although not systematically, triglyceride hepatic concentration) in rats or hamsters (Roberfroid et al. 2010).

Improvement of glycemia has also been observed in healthy adults, but not in diabetic patients (Cani et al. 2009; Roberfroid et al. 2010). Fructans decrease the plasma concentration of low density lipoprotein (LDL) cholesterol and the LDL to high density lipoprotein (HDL) ratio in hyperlipidemic adults, but do not seem to induce any particular change in normolipidemic individuals (Roberfroid et al. 2010; de Luis et al. 2011; Brownawell et al. 2012). Fructans also decrease the hepatic capacity of triglyceride synthesis in human volunteers (Roberfroid et al. 2010).

Despite some discrepancy (Boucher et al. 2003), GOS and a GOS and inulin mix also induce decreases in cholesterol in human microbiota associated rats (Djouzi and Andrieux 1997), in total cholesterol and LDL levels in infants (Alliet et al. 2007) or in total cholesterol (TC), triglycerides, and the TC to HDL cholesterol ratio in obese adults (Vulevic et al. 2013).

Again numerous mechanisms are supposedly involved in these physiological effects: improvement of glycemia could stem from stimulation of GLP-1 production (Delzenne 2003). A decrease in de novo hepatic lipogenesis could be the result of the enhancement of SCFA, particularly propionate, which is reported to inhibit fatty acid synthesis in vitro (Delzenne and Kok 2001), but also acetate, which also exerts an inhibitory effect on lipogenesis, despite the fact that is a precursor for lipogenesis (Roberfroid et al. 2010).

#### 9.5.1.6 Modulation of Immune Functions

The immuno-modulatory potential of prebiotics has motivated a few studies. As a reflection of the complexity of the immune response (innate vs adaptive, mucosal vs systemic, pro-inflammatory vs anti-inflammatory), numerous markers have been considered (vaccine-specific serum antibody production, delayed-type hypersensitivity response, vaccine-specific or total secretory immunoglobulin A (sIgA) in saliva, the response to attenuated pathogens, NK cell activity, phagocytosis, T-cell proliferation, cytokine concentrations) (Roberfroid et al. 2010).

Several studies have demonstrated an increased intestinal sIgA response, an increase in B cell numbers in Peyer's patches, and an enhanced intestinal interleukin 10 (IL-10) protein secretion in intestinal tissues, as well as a decreased mRNA expression and protein concentration of pro-inflammatory cytokines resulting from the use of FOS in animal models (Macfarlane et al. 2008; Roberfroid et al. 2010). Furthermore, functional activities of Natural Killer (NK) cells and phagocytes isolated from various immune tissues were significantly increased, but depending on the source of immune cells (Peyer's patches, mesenteric lymph nodes, intraepithelial lymphocytes) the prebiotic effects may differ (Roberfroid et al. 2010).

In humans, where infants and elderly were mainly investigated, fructans on their own seem poorly effective in stimulating a response to vaccination, while a mix of GOS and inulin (or even with a stronger response, a mix of GOS, inulin and AOS (acidic oligosaccharides derived from pectin)) appears effective in enhancing T helper cells type 1 ($T_h1$) responsiveness (Jeurink et al. 2013). A GOS and inulin mix also enhanced faecal sIgA concentrations in infants (Roberfroid et al. 2010).

A similar impact of prebiotics seems to occur in older people: FOS induced a decrease in phagocytosis and IL-6 mRNA expression in peripheral blood mononuclear cells in healthy elderly people (Guigoz et al. 2002). However, an inulin and FOS mix had no impact on sIgA, serum titers after vaccination (influenza A and B and *Pneumococcus*), secretion of IL-4 and interferon gamma (IFN-γ) or on lymphocyte proliferation (Bunout et al. 2002). GOS consumption, on the other hand, resulted in increases in ex vivo NK cell activity, ex vivo phagocytosis and ex vivo IL-10 production by peripheral blood mononuclear cells (PBMC) together with a decrease in ex vivo IL-6, tumor necrosis factor alpha (TNFα) and IL-1β production by PBMC (Vulevic et al. 2008).

Prebiotic effects may impact the immune response indirectly as a result of intestinal fermentation and promotion of growth of certain members of the gut microbiota (Macfarlane et al. 2008; Roberfroid et al. 2010). SCFA have direct immunomodulatory properties. G-protein-coupled receptors (GPR41 and GPR43), which have been identified as receptors for SCFA, are expressed on leukocytes, as well as on enterocytes and enteroendocrine cells in the human colon (Roberfroid et al. 2010). Interestingly, GPR43-deficient (Gpr43(−/−)) mice showed exacerbated inflammatory responses to various pathological situations (Maslowski et al. 2009).

Butyrate modulates chemokine expression in intestinal epithelial cells, differentially affects pro-inflammatory IL-2, IFN-γ and immunoregulatory IL-10 production by rat lymphocytes in vitro, through regulation of the transcription factor NF-κB (Macfarlane et al. 2008; Roberfroid et al. 2010). Acetate increases peripheral blood antibody production and NK activity when administrated intravenously (Macfarlane et al. 2008). Lastly, SCFA could improve the immune response through an indirect effect: because they are used by epithelial cells as energy substrates, their increased production allows for the sparing of glutamine. As glutamine is used by immune cells in the body, this thereby enhances immune system reactivity (Jenkins et al. 1999).

Both commensal and exogenous bacteria interact with both the innate and adaptive immune system through pattern-recognition receptors, such as the toll-like receptors (TLR), in a strain-dependent manner. Thus, the presence of increased numbers of a particular microbial genus or species, or a related decrease of other microbes, may change the collective immuno-interactive profile of the microbiota, thus resulting in a variety of downstream events. This may eventually lead to cytokine production steering towards an appropriate immune response for the microbial event (Roberfroid et al. 2010). For example, some lactobacilli can enhance both the innate and adaptive immune defences, resulting in increased production of phagocytes and immune effector molecules such as sIgA, and some bifidobacteria are able to induce formation of large amounts of IgA in Peyer's patches of mice (Macfarlane et al. 2008). As a possible illustration of such potential, a positive correlation between numbers of bifidobacteria in faecal samples and both NK cell activity and phagocytosis was observed in the study of Vulevic et al. (2008) mentioned above. However, lactobacilli and bifidobacteria are far from being the sole bacteria showing this potential, as illustrated by recent studies which have highlighted, for example, the immunomodulatory effects of other bacteria such segmented filamentous bacteria (Gaboriau-Routhiau et al. 2009) or *F. prausnitzii* (Sokol et al. 2008).

Finally, it has been postulated that prebiotics could interact directly with epithelial cells by specific lectin-like receptors (Seifert and Watzl 2007). This remains only theoretical today since the existence of specific receptors for FOS has not been demonstrated to date, although some indirect evidence obtained in vitro argues in its favour, such as the inhibition of phagocytosis of bacteria by granulocytes due to the addition of fructose (Speert et al. 1984) or the activation of NK activity of peripheral blood mononuclear cells due to the addition of a plant extract containing FOS (Thakur et al. 2012). It has been noticed that if such a mechanism is proven true, this will challenge the prebiotic definition, which supposes a causal relation between the physiological effect of the prebiotic and its impact on the composition and/or activity of the gastrointestinal microbiota.

### 9.5.2 Current Evidence for Disease Prevention or Treatment

Whether the physiological effects outlined above actually translate into reduction of the risk and/or treatment possibilities of diseases is not established yet. Prebiotics appear effective for the treatment of infectious diarrhea (Vandenplas et al. 2013), irritable bowel syndrome (Whelan 2011) and for the prevention of eczema in infants (Osborn and Sinn 2013). For most other cases (gut inflammatory diseases, colorectal cancer, metabolic diseases, osteoporosis, other allergies) there are promising results from preclinical studies (i.e., in animal models of disease) or from small pilot studies, but appropriate human intervention studies are too scarce to conclude on the real impact of prebiotics (Roberfroid et al. 2010; Brownawell et al. 2012). For example, although prebiotics such as FOS, inulin and GOS appear to favourably alter biomarkers of cardiovascular diseases, including low-density lipoprotein cholesterol (LDL-C) (see above) and C-reactive protein (CRP) (De Luis et al. 2011; Dehghan et al. 2014; Vulevic et al. 2013), whether they actually decrease the risk of CVD is unclear as yet (Brownawell et al. 2012; Slavin 2013).

Similarly, prebiotic intervention decreases numerous hallmarks of obesity and diabetes such as food intake, fat storage in adipose tissue and in the liver (steatosis), glycemia and hepatic insulin resistance, endotoxemia and systemic inflammation, in several models of obese rodents (Delzenne and Cani 2011; Everard and Cani 2013). However, from the limited number of intervention studies which have investigated the role of fructans in humans, the actual weight loss (a few kg) in obese individuals supplemented with fructans remains modest (Delzenne et al. 2013) or not statistically significant (Dewulf et al. 2013). Furthermore, no long-term benefit on diabetes has been demonstrated yet (Everard and Cani 2013) and preliminary results from acute impact are contradictory (Dehghan et al. 2014; Dewulf et al. 2013).

Also with respect to osteoporosis, it is too early to establish whether the beneficial impact of fructans on calcium metabolism actually translates into reduction of the risk of bone disease, even if bone-sparing effects have been found in ovariectomized rats (an animal model for osteoporosis) (Coxam 2007). Such a risk reduction would require proof of the persistance of the benefits of prebiotics on calcium absorption (Roberfroid et al. 2010). In this respect, the fact that improvement of calcium metabolism was still detectable after 1 year (Abrams et al. 2005) and that inulin-based fibres exert chronic effects on calcium utilization in a postmenopausal rodent model (Legette et al. 2012) appear promising, and prebiotic-induced improvement of calcium metabolism may have important implications for future preventative strategies for osteoporosis.

The diseases covered here are not an exhaustive list, as it becomes ever clearer that the actions of the microbiota have far-reaching consequences. For example, there appears to be a link between the microbiota and the nervous system, and psychiatric diseases may become a target for therapy via microbiota modulation (for a recent review see Collins et al. 2012).

## 9.6 Conclusions

We have seen a vast expansion of our knowledge of the microbial community resident in the intestine in recent years, and it is becoming clear that the prebiotic concept as it currently stands may be too simplistic in view of the complexity of the biological system prebiotics are supposed to modulate. Furthermore, it is difficult to make a clear distinction between microbiota effects mediated via fermentation of prebiotics or via other microbial activities, such as biotransformation of secondary plant products to more health-promoting derivatives. This is further complicated by the fact that prebiotics may also exert health-promoting effects that are independent of the microbiota. Thus further mechanistic work is necessary to prove a casual relationship between microbiota changes and health effects, as a correlation between the two may be by mere association.

In view of recent work on the potential importance of gut microbial diversity in health and the differences in microbiota composition observed between individuals, further work is also urgently needed on different study groups as well as different individuals. A study on the changes in the butyrate-producing community indicated that individuals with the most common bacterial profile showed little change in response to inulin intake, whereas three individuals with very unique profiles showed major changes (Louis et al. 2010). While this will have to be confirmed by larger studies, it indicates that some individuals may benefit more from prebiotic intake than others. Some evidence also exists that prebiotic intake may actually be harmful for specific conditions, such as certain gut infections (Licht et al. 2012). When considering overall microbiota diversity, a diet rich in a variety of different fibres may actually be more beneficial for some individuals or disease states. As different fibres are likely to exert different health effects, this may lead to a better overall improvement in health and disease prevention (Slavin 2013; Raninen et al. 2011). Nevertheless, in view of the recent evidence of how far beyond the intestine microbial activties are exerting their effects, targeting specific changes in the microbiota looks like an increasingly promising concept to improve human health.

## References


[No authors listed] (2005) Br J Nutr 93(Suppl 1):S1–168

[No authors listed] (2007) Inulin and oligofructose: proven health benefits and claims. Proceedings of the 5th ORAFTI Research Conference, 28–29 Sept 2006, Boston, MA. J Nutr 137(11 Suppl):2489S–2597S

Abrams SA, Griffin IJ, Hawthorne KM, Liang L, Gunn SK, Darlington G, Ellis KJ (2005) A combination of prebiotic short- and long-chain inulin-type fructans enhances calcium absorption and bone mineralization in young adolescents. Am J Clin Nutr 82:471–476

Alliet P, Scholtens P, Raes M, Hensen K, Jongen H, Rummens JL, Boehm G, Vandenplas Y (2007) Effect of prebiotic galacto-oligosaccharide, long-chain fructo-oligosaccharide infant formula on serum cholesterol and triacylglycerol levels. Nutrition 23:719–723

Anastasovska J, Arora T, Sanchez Canon GJ, Parkinson JR, Touhy K, Gibson GR, Nadkarni NA, So PW, Goldstone AP, Thomas EL, Hankir MK, Van Loo J, Modi N, Bell JD, Frost G (2012) Fermentable carbohydrate alters hypothalamic neuronal activity and protects against the obesogenic environment. Obesity (Silver Spring) 20:1016–1023

Ashley C, Johnston WH, Harris CL, Stolz SI, Wampler JL, Berseth CL (2012) Growth and tolerance of infants fed formula supplemented with polydextrose (PDX) and/or galactooligosaccharides (GOS): double-blind, randomized, controlled trial. Nutr J 11:38

Barrat E, Michel C, Poupeau G, David-Sochard A, Rival M, Pagniez A, Champ M, Darmaun D (2008)

Supplementation with galactooligosaccharides and inulin increases bacterial translocation in artificially reared newborn rats. Pediatr Res 64:34–39

Belenguer A, Duncan SH, Calder G, Holtrop G, Louis P, Lobley GE, Flint HJ (2006) Two routes of metabolic cross-feeding between *Bifidobacterium adolescentis* and butyrate-producing anaerobes from the human gut. Appl Environ Microbiol 72:3593–3599

Belenguer A, Duncan SH, Holtrop G, Anderson S, Lobley GE, Flint HJ (2007) Impact of pH on lactate formation and utilisation by human fecal microbial communities. Appl Environ Microbiol 73:6526–6533

Bergillos-Meca T, Navarro-Alarcón M, Cabrera-Vique C, Artacho R, Olalla M, Giménez R, Moreno-Montoro M, Ruiz-Bravo A, Lasserrot A, Ruiz-López MD (2013) The probiotic bacterial strain *Lactobacillus fermentum* D3 increases in vitro the bioavailability of Ca, P, and Zn in fermented goat milk. Biol Trace Elem Res 151:307–314

Boucher J, Daviaud D, Siméon-Remaud M, Carpéné C, Saulnier-Blache JS, Monsan P, Valet P (2003) Effect of non-digestible gluco-oligosaccharides on glucose sensitivity in high fat diet fed mice. J Physiol Biochem 59:169–173

Bouhnik Y, Raskine L, Simoneau G, Vicaut E, Neut C, Flourié B, Brouns F, Bornet FR (2004) The capacity of nondigestible carbohydrates to stimulate faecal bifidobacteria in healthy humans: a double blind, randomized, placebo-controlled, parallel-group, dose response relation study. Am J Clin Nutr 80:1658–1664

Bourriaud C, Robins RJ, Martin L, Kozlowski F, Tenailleau E, Cherbut C, Michel C (2005) Lactate is mainly fermented to butyrate by human intestinal microfloras but inter-individual variation is evident. J Appl Microbiol 99:201–212

Brownawell AM, Caers W, Gibson GR, Kendall CW, Lewis KD, Ringel Y, Slavin JL (2012) Prebiotics and the health benefits of fiber: current regulatory status, future research, and goals. J Nutr 142:962–974

Bunout D, Hirsch S, de la Maza MP, Munoz C, Hascke F, Steenhout P, Klassen P, Barrera G, Gattas V, Petermann M (2002) Effects of prebiotics on the immune response to vaccination in the elderly. JPEN Parenter Enter 26:372–376

Candela M, Maccaferri S, Turroni S, Carnevali P, Brigidi P (2010) Functional intestinal microbiome, new frontiers in prebiotic design. Int J Food Microbiol 140:93–101

Cani PD, Knauf C, Iglesias MA, Drucker DJ, Delzenne NM, Burcelin R (2006a) Improvement of glucose tolerance and hepatic insulin sensitivity by oligofructose requires a functional glucagon-like peptide 1 receptor. Diabetes 55:1484–1490

Cani PD, Joly E, Horsmans Y, Delzenne NM (2006b) Oligofructose promotes satiety in healthy human: a pilot study. Eur J Clin Nutr 60:567–572

Cani PD, Possemiers S, Van de Wiele T, Guiot Y, Everard A, Rottier O, Geurts L, Naslain D, Neyrinck A, Lambert DM, Muccioli GG, Delzenne NM (2009) Changes in gut microbiota control inflammation in obese mice through a mechanism involving GLP-2-driven improvement of gut permeability. Gut 58:1091–1103

Cani PD, Osto M, Geurts L, Everard A (2012) Involvement of gut microbiota in the development of low-grade inflammation and type 2 diabetes associated with obesity. Gut Microbes 3:279–288

Cecchini DA, Laville E, Laguerre S, Robe P, Leclerc M, Doré J, Henrissat B, Remaud-Siméon M, Monsan P, Potocki-Véronèse G (2013) Functional metagenomics reveals novel pathways of prebiotic breakdown by human gut bacteria. PLoS ONE 8:e72766

Chung WSF, Walker AW, Louis P, Parkhill J, Vermeiren J, Bosscher D, Duncan SH, Flint HJ (2016) Modulation of the human gut microbiota by dietary fibres occurs at the species level. BMC Biol 14:3

Collins SM, Surette M, Bercik P (2012) The interplay between the intestinal microbiota and the brain. Nat Rev Microbiol 10:735–742

Costabile A, Fava F, Röytiö H, Forssten SD, Olli K, Klievink J, Rowland IR, Ouwehand AC, Rastall RA, Gibson GR, Walton GE (2012) Impact of polydextrose on the faecal microbiota: a double-blind, crossover, placebo-controlled feeding study in health human subjects. Br J Nutr 108:471–481

Cotillard A, Kennedy SP, Kong LC, Prifti E, Pons N, Le Chatelier E, Almeida M, Qunquis B, Levenez F, Galleron N, Gougis S, Rizkalla S, Batto J-M, Renault P, ANR MicroObes consortium, Doré J, Zucker J-D, Clément K, Ehrlich SD (2013) Dietary intervention impact on gut microbial gene richness. Nature 500:585–588

Coxam V (2007) Current data with inulin-type fructans and calcium, targeting bone health in adults. J Nutr 137:2527S–2533S

Cummings JH, Macfarlane GT (2002) Gastrointestinal effects of prebiotics. Br J Nutr 87:S145–S151

Dalvi PS, Nazarians-Armavil A, Purser MJ, Belsham D (2012) Glucagon-like peptide-1 receptor agonist, exendin-4, regulates feeding-associated neuropeptides in hypothalamic neurons in vivo and in vitro. Endocrinology 153:2208–2222

de Luis DA, de la Fuente B, Izaola O, Conde R, Gutiérrez S, Morillo M, Teba Torres C (2011) Double blind randomized clinical trial controlled by placebo with an alpha linoleic acid and prebiotic enriched cookie on risk cardiovascular factor in obese patients. Nutr Hosp 26:827–833

Dehghan P, Gargari BP, Jafar-Abadi MA, Aliasgharzadeh A (2014) Inulin controls inflammation and metabolic endotoxemia in women with type 2 diabetes mellitus: a randomized-controlled clinical trial. Int J Food Sci Nutr 65(1):117–123

Delzenne N (2003) Oligosaccharides: state of the art. Proc Nutr Soc 62:177–182

Delzenne NM, Cani PD (2011) Interaction between obesity and the gut microbiota: relevance in nutrition. Annu Rev Nutr 31:15–31

Delzenne NM, Kok N (2001) Effects of fructans-type prebiotics on lipid metabolism. Am J Clin Nutr 73:456S–458S

Delzenne NM, Neyrinck AM, Cani PD (2013) Gut microbiota and metabolic disorders: how prebiotic can work? Br J Nutr 109:S81–S85

Dewulf EM, Cani P, Claus SP, Fuentes S, Puylaert PGB, Neyrinck AM, Bindels LB, de Vos WM, Gibson GR, Thissen J-P, Delzenne NM (2013) Insight into the prebiotic concept: lessons from an exploratory, double blind intervention study with inulin-type fructans in obese women. Gut 62:1112–1121

Djouzi Z, Andrieux C (1997) Compared effects of three oligosaccharides on metabolism of intestinal microflora in rats inoculated with a human faecal flora. Br J Nutr 78:313–324

Duncan SH, Hold GL, Harmsen HJM, Stewart CS, Flint HJ (2002) Growth requirements and fermentation products of *Fusobacterium prausnitzii*, and a proposal to reclassify it as *Faecalibacterium prausnitzii* gen. nov., comb. nov. Int J Syst Evol Microbiol 52:2141–2146

Duncan SH, Louis P, Flint HJ (2004) Lactate-utilizing bacteria, isolated from human feces, that produce butyrate as a major fermentation product. Appl Environ Microbiol 70:5810–5817

Duncan SH, Louis P, Thomson JM, Flint HJ (2009) The role of pH in determining the species composition of the human colonic microbiota. Environ Microbiol 11:2112–2122

EFSA Panel on Dietatic Products Nutrition and Allergies (2011) Guidance on the scientific requirements for health claims related to gut and immune function. EFSA J 9:1984

Everard A, Cani PD (2013) Diabetes, obesity and gut microbiota. Best Pract Res Clin Gastroenterol 27:73–83

Ewaschuk JB, Naylor JM, Zello GA (2005) D-lactate in human and ruminant metabolism. J Nutr 135:1619–1625

Falony G, Vlachou A, Verbrugghe K, de Vuyst L (2006) Cross-feeding between *Bifidobacterium longum* BB536 and acetate-converting, butyrate-producing colon bacteria during growth on oligofructose. Appl Environ Microbiol 72:7835–7841

Flint HJ, Duncan SH, Scott KP, Louis P (2007) Interactions and competition within the microbial community of the human colon: links between diet and health. Environ Microbiol 9:1101–1111

Flint HJ, Scott KP, Louis P, Duncan SH (2012a) The role of the gut microbiota in nutrition and health. Nat Rev Gastroenterol Hepatol 9:577–589

Flint HJ, Scott KP, Duncan SH, Louis P, Forano E (2012b) Microbial degradation of complex carbohydrates in the gut. Gut Microbes 3:289–306

Fuentes-Zaragoza E, Sánchez-Zapata E, Sendra E, Sayas E, Navarro C, Fernández-López J, Pérez-Alvarez JA (2011) Resistant starch as prebiotic: a review. Starch 63:406–415

Fukushima A, Aizaki Y, Sakuma K (2012) Short-chain fatty acids increase the level of calbindin-D9k messenger RNA in Caco-2 cells. J Nutr Sci Vitaminol (Tokyo) 58:287–291

Gaboriau-Routhiau V, Rakotobe S, Lécuyer E, Mulder I, Lan A, Bridonneau C, Rochet V, Pisi A, De Paepe M, Brandi G, Eberl G, Snel J, Kelly D, Cerf-Bensussan N (2009) The key role of segmented filamentous bacteria in the coordinated maturation of gut helper T cell responses. Immunity 31:677–689

Gibson GR, Roberfroid MB (1995) Dietary modulation of the human colonic microbiota: introducing the concept of prebiotics. J Nutr 125:1401–1412

Gibson GR, Scott KP, Rastall RA, Tuohy KM, Hotchkiss A, Dubert-Ferrandon A, Gareau M, Murphy EF, Saulnier D, Loh G, Macfarlane S, Delzenne N, Ringel Y, Kozianowski G, Dickmann R, Lenoir-Wijnkoop I, Walker C, Buddington R (2010) Dietary prebiotics: current status and new definition. Food Sci Technol Bull Funct Foods 7:1–19

Gilman J, Cashman KD (2006) The effect of probiotic bacteria on transepithelial calcium transport and calcium uptake in human intestinal-like Caco-2 cells. Curr Issues Intest Microbiol 7:1–5

Goffin D, Delzenne N, Blecker C, Hanon E, Deroanne C, Paquot M (2011) Will isomalto-oligosaccharides, a well-established functional food in Asia, break through the European and American market? The status of knowledge on these prebiotics. Crit Rev Food Sci Nutr 51:394–409

Griffin IJ, Davila PM, Abrams SA (2002) Non-digestible oligosaccharides and calcium absorption in girls with adequate calcium intakes. Br J Nutr 87:S187–S191

Guigoz Y, Rochat F, Perruisseau-Carrier G, Rochat I, Schiffrin EJ (2002) Effects of oligosaccharide on the faecal flora and non-specific immune system in elderly people. Nutr Res 22:13–25

Gullón B, Gómez B, Martínez-Sabajanes M, Yánez R, Parajó JC, Alonso JL (2013) Pectic oligosaccharides: manufacture and functional properties. Trends Food Sci Technol 30:153–161

Hamer HM, Jonkers D, Venema K, Vanhoutvin S, Troost FJ, Brummer RJ (2008) Review article: the role of butyrate on colonic function. Aliment Pharmacol Ther 27:104–119

Hess JR, Birkett AM, Thomas W, Slavin JL (2011) Effects of short-chain fructooligosaccharides on satiety responses in healthy men and women. Appetite 56:128–134

Hicks PD, Hawthorne KM, Berseth CL, Marunycz JD, Heubi JE, Abrams SA (2012) Total calcium absorption is similar from infant formulas with and without prebiotics and exceeds that in human milk-fed infants. BMC Pediatr 12:118

Howlett JF, Betteridge VA, Champ M, Craig SAS, Meheust A, Jones JM (2010) The definition of dietary fiber – discussions at the Ninth Vahouny Fiber Symposium: building scientific agreement. Food Nutr Res 54:5750

Jenkins DJ, Kendall CW, Vuksan V (1999) Inulin, oligofructose and intestinal function. J Nutr 129:1431S–1433S

Jeurink PV, van Esch BC, Rijnierse A, Garssen J, Knippels LM (2013) Mechanisms underlying immune effects of dietary oligosaccharides. Am J Clin Nutr 98:572S–577S

Johnson CR, Thavarajah D, Combs GF Jr, Thavarajah P (2013) Lentil (*Lens culinaris* L.): a prebiotic-rich whole food legume. Food Res Int 51:107–113

Kleessen B, Blaut M (2005) Modulation of gut mucosal biofilms. Br J Nutr 93:S35–S40

Kruger MC, Brown KE, Collett G, Layton L, Schollum LM (2003) The effect of fructooligosaccharides with various degrees of polymerization on calcium bioavailability in the growing rat. Exp Biol Med (Maywood) 228:683–688

Le Chatelier E, Nielsen T, Qin J, Prifti E, Hildebrand F, Falony G, Ameida M, Arumugam M, Batto J-M, Kennedy S, Leonard P, Li J, Burgdorf K, Grarup N, Jørgensen T, Brandslund I, Nielsen HB, Juncker AS, Bertalan M, Levenez F, Pons N, Rasmussen S, Sunagawa S, Tap J, Tims S, Zoetendal EG, Brunak S, Clément K, Doré J, Kleerebezem M, Kristiansen K, Renault P, Sicheritz-Ponten T, de Vos WM, Zucker J-D, Raes J, Hansen T, MetaHIT consortium, Bork P, Wang J, Ehrlich DS, Pedersen O (2013) Richness of human gut microbiome correlates woth metabolic markers. Nature 500:541–546

Legette LL, Lee W, Martin BR, Story JA, Campbell JK, Weaver CM (2012) Prebiotics enhance magnesium absorption and inulin-based fibers exert chronic effects on calcium utilization in a postmenopausal rodent model. J Food Sci 77:H88–H94

Licht TR, Ebersbach T, Frøkiær H (2012) Prebiotics for prevention of gut infections. Trends Food Sci Technol 23:70–82

Lopez-Siles M, Khan TM, Duncan SH, Harmsen HM, Garcia-Gil LJ, Flint HJ (2012) Cultured representatives of two major phylogroups of human colonic *Faecalibacterium prausnitzii* can utilize pectin, uronic acids, and host-derived substrates for growth. Appl Environ Microbiol 78:420–428

Louis P, Flint HJ (2009) Diversity, metabolism and microbial ecology of butyrate-producing bacteria from the human large intestine. FEMS Microbiol Lett 294:1–8

Louis P, O'Byrne CP (2010) Life in the gut: microbial responses to stress in the gastrointestinal tract. Sci Prog 93:7–36

Louis P, Young P, Holtrop G, Flint HJ (2010) Diversity of human colonic butyrate-producing bacteria revealed by analysis of the butyryl-CoA: acetate CoA-transferase gene. Environ Microbiol 12:304–314

Lozupone CA, Stornbaugh JI, Gordon JI, Jansson JK, Knight R (2012) Diversity, stability and resilience of the human gut microbiota. Nature 489:220–230

Maathuis AJH, van den Heuvel EG, Schoterman MHC, Venema K (2012) Galacto-oligosaccharides have prebiotic activity in a dynamic in vitro colon model using a 13C-labeling technique. J Nutr 142:1205–1212

Macfarlane GT, Macfarlane S (2012) Bacteria, colonic fermentation, and gastrointestinal health. J AOAC Int 95:50–60

Macfarlane GT, Steed H, Macfarlane S (2008) Bacterial metabolism and health-related effects of galacto-oligosaccharides and other prebiotics. J Appl Microbiol 104:305–344

Marín-Manzano MC, Abecia L, Hernández-Hernández O, Sanz ML, Montilla A, Olano A, Rubio LA, Moreno FJ, Clemente A (2013) Galacto-oligosaccharides derived from lactulose exert a selective stimulation on the growth of *Bifidobacterium animalis* in the large intestine of growing rats. J Agric Food Chem 61:7560–7567

Martens EC, Koropatkin NM, Smith TJ, Gordon JI (2009) Complex glycan catabolism by the human gut microbiota: the Bacteroidetes sus-like paradigm. J Biol Chem 284:24673–24677

Maslowski KM, Vieira AT, Ng A, Kranich J, Sierro F, Yu D, Schilter HC, Rolph MS, Mackay F, Artis D, Xavier RJ, Teixeira MM, Mackay CR (2009) Regulation of inflammatory responses by gut microbiota and chemoattractant receptor GPR43. Nature 2461:1282–1286

McIntosh FM, Maison N, Holtrop G, Young P, Stevens VH, Ince J, Johnstone AM, Lobley GE, Flint HJ, Louis P (2012) Phylogenetic distribution of genes encoding β-glucuronidase activity in human colonic bacteria and the impact of diet on faecal glycosidase activities. Environ Microbiol 14:1876–1887

Meslin JC, Andrieux C, Sakata T, Beaumatin P, Bensaada M, Popot F, Szylit O, Durand M (1993) Effects of galacto-oligosaccharide and bacterial status on mucin distribution in mucosa and on large intestine fermentation in rats. Br J Nutr 69:903–912

Moro GE, Stahl B, Fanaro S, Jelinek J, Boehm G, Coppa GV (2005) Dietary prebiotic oligosaccharides are detectable in the faeces of formula-fed infants. Acta Paediatr Suppl 94:27–30

Morrison DJ, Mackay WG, Edwards CA, Preston T, Dodson B, Weaver LT (2006) Butyrate production from oligofructose fermentation by the human faecal flora: what is the contribution of extracellular acetate and lactate? Br J Nutr 96:570–577

Neyrinck AM, Van Hée VF, Piront N, De Backer F, Toussaint O, Cani PD, Delzenne NM (2012) Wheat-derived arabinoxylan oligosaccharides with prebiotic effect increase satietogenic gut peptides and reduce metabolic endotoxemia in diet-induced obese mice. Nutr Diabetes 2:e28

Ohland CL, Macnaughton WK (2010) Probiotic bacteria and intestinal epithelial barrier function. Am J Physiol Gastrointest Liver Physiol 298:G807–G819

Ohta A, Motohashi Y, Ohtsuki M, Hirayama M, Adachi T, Sakuma K (1998) Dietary fructooligosaccharides change the concentration of calbindin-D9k differently in the mucosa of the small and large intestine of rats. J Nutr 128:934–939

Osborn DA, Sinn JK (2013) Prebiotics in infants for prevention of allergy. Cochrane Database Syst Rev 3:CD006474

Otieno DO, Ahring BK (2012) The potential for oligosaccharide production from the hemicellulose fraction of biomasses through pretreatment processes: xylooligosaccharides (XOS), arabinooligosaccharides (AOS), and mannooligosaccharides (MOS). Carbohydr Res 360:84–92

Overduin J, Schoterman MH, Calame W, Schonewille AJ, Ten Bruggencate SJ (2013) Dietary galacto-oligosaccharides and calcium: effects on energy intake, fat-pad weight and satiety-related, gastrointestinal hormones in rats. Br J Nutr 109:1338–1348

Parnell JA, Reimer RA (2009) Weight loss during oligofructose supplementation is associated with decreased ghrelin and increased peptide YY in overweight and obese adults. Am J Clin Nutr 89:1751–1759

Peters HP, Boers HM, Haddeman E, Melnikov SM, Qvyjt F (2009) No effect of added beta-glucan or of fructooligosaccharide on appetite or energy intake. Am J Clin Nutr 89:58–63

Ramirez-Farias C, Slezak K, Fuller Z, Duncan A, Holtrop G, Louis P (2009) Effect of inulin on the human gut microbiota: stimulation of *Bifidobacterium adolescentis* and *Faecalibacterium prausnitzii*. Br J Nutr 101:541–550

Raninen K, Lappi J, Mykkänen H, Poutanen K (2011) Dietary fiber type reflects physiological functionality: comparison of grain fiber, inulin, and polydextrose. Nutr Rev 69:9–21

Roberfroid M, Gibson GR, Hoyles L, McCartney AL, Rastall R, Rowland I, Wolvers D, Watzl B, Szajewska H, Stahl B, Guarner F, Respondek F, Whelan K, Coxam V, Davicco MJ, Léotoing L, Wittrant Y, Delzenne NM, Cani PD, Neyrinck AM, Meheust A (2010) Prebiotic effects: metabolic and health benefits. Br J Nutr 104:S1–S63

Rossi M, Corradini C, Amaretti A, Nicolini M, Pompei A, Zanoni S, Matteuzzi D (2005) Fermentation of fructooligosaccharides and inulin by bifidobacteria: a comparative study of pure and fecal cultures. Appl Environ Microbiol 71:6150–6158

Rozan P, Nejdi A, Hidalgo S, Bisson JF, Desor D, Messaoudi M (2008) Effects of lifelong intervention with an oligofructose-enriched inulin in rats on general health and lifespan. Br J Nutr 100:1192–1199

Russell W, Duthie G (2011) Symposium on 'nutrition: getting the balance right in 2010'. Session 3: influences of food constituents on gut health plant secondary metabolites and gut health: the case for phenolic acids. Proc Nutr Soc 70:389–396

Russell WR, Hoyles L, Flint HJ, Dumas ME (2013) Colonic bacterial metabolites and human health. Curr Opin Microbiol 16:246–254

Russo F, Linsalata M, Clemente C, Chiloiro M, Orlando A, Marconi E, Chimienti G, Riezzo G (2012) Inulin-enriched pasta improves intestinal permeability and modifies the circulating levels of zonulin and glucagon-like peptide 2 in healthy young volunteers. Nutr Res 32:940–946

Ryan SM, Fitzgerald GF, Van Sinderen D (2006) Screening for and identification of starch-, amylopectin-, and pullulan-degrading activities in Bifidobacterial strains. Appl Environ Microbiol 72:5289–5296

Scholz-Ahrens KE, Schrezenmeir J (2002) Inulin, oligofructose and mineral metabolism – experimental data and mechanism. Br J Nutr 87:S179–S186

Scholz-Ahrens KE, Ade P, Marten B, Weber P, Timm W, Açil Y, Glüer CC, Schrezenmeir J (2007) Prebiotics, probiotics, and synbiotics affect mineral absorption, bone mineral content, and bone structure. J Nutr 137:838S–846S

Scott KP, Martin JC, Duncan SH, Flint HJ (2013) Prebiotic stimulation of human colonic butyrate-producing bacteria and bifidobacteria, *in vitro*. FEMS Microbiol Ecol 87(1):30–40

Seifert S, Watzl B (2007) Inulin and oligofructose: review of experimental data on immune modulation. J Nutr 137:2563S–2567S

Slavin J (2013) Fiber and prebiotics: mechanisms and health benefits. Nutrients 5:1417–1435

Slavin J, Green H (2007) Dietary fibre and satiety. Nutr Bull 32:32–42

Sokol H, Pigneur B, Watterlot L, Lakhdari O, Bermúdez-Humarán LG, Gratadoux J-J, Blugeon S, Bridonneau C, Furet J-P, Corthier G, Grangette C, Vasquez N, Pochart P, Trugnan G, Thomas G, Blottière HM, Doré J, Marteau P, Seksik P, Langella P (2008) *Faecalibacterium prausnitzii* is an anti-inflammatory commensal bacterium identified by gut microbiota analysis of Crohn disease patients. Proc Natl Acad Sci U S A 105:16731–16736

Salonen A, Lahti L, Salojärvi J, Holtrop G, Korpela K, Duncan SH, Date P, Farquharson F, Johnstone AM, Lobley GE, Louis P, Flint HJ, de Vos W (2014) Impact of diet and individual variation on intestinal microbiota composition and fermentation products in obese men. ISME J 8:2218–2230

Speert DP, Eftekhar F, Puterman ML (1984) Nonopsonic phagocytosis of strains of *Pseudomonas aeruginosa* from cystic fibrosis patients. Infect Immun 43:1006–1011

Tabbers MM, Boluyt N, Berger MY, Benninga MA (2011) Nonpharmacologic treatments for childhood constipation: systematic review. Pediatrics 128:753–761

Tap J, Mondot S, Levenez F, Pelletier E, Caron C, Furet J-P, Ugarte E, Munoz-Tamayo R, Paslier DLE, Nallin R, Doré J, Leclerc M (2009) Towards the human intestinal microbiota phylogenetic core. Environ Microbiol 11:2574–2584

Thakur M, Connellan P, Deseo MA, Morris C, Praznik W, Loeppert R, Dixit VK (2012) Characterization and in vitro immunomodulatory screening of fructo-oligosaccharides of Asparagus racemosus Willd. Int J Biol Macromol 50:77–81

Tolhurst G, Heffron H, Lam YS, Parker HE, Habib AM, Diakogiannaki E, Cameron J, Grosse J, Reimann F,

Gribble F (2012) Short-chain fatty acids stimulate glucagon-like peptide-1 secretion via the G-protein-coupled receptor FFAR2. Diabetes 61:364–371

Tzounis X, Rodriguez-Mateos A, Vulevic J, Gibson GR, Kwik-Uribe C, Spencer JPE (2011) Prebiotic evaluation of cocoa-derived flavanols in healthy humans by using a randomized, controlled, double-blind, crossover intervention study. Am J Clin Nutr 93:62–72

Van den Abbeele P, Verstraete W, El Aidy S, Geirnaert A, Van de Wiele T (2013) Prebiotics, faecal transplants and microbial network units to stimulate biodiversity of the human gut microbiome. Microb Biotechnol 6:335–340

van den Heuvel EG, Schoterman MH, Muijs T (2000) Transgalactooligosaccharides stimulate calcium absorption in postmenopausal women. J Nutr 130:2938–2942

Vandenplas Y, De Greef E, Hauser B, Devreker T, Veereman-Wauters G (2013) Probiotics and prebiotics in pediatric diarrheal disorders. Expert Opin Pharmacother 14:397–409

Vernia P, Caprilli R, Latella G, Barbetti F, Magliocca FM, Cittadini M (1988) Fecal lactate and ulcerative colitis. Gastroenterology 95:1564–1568

Vulevic J, Drakoularakou A, Yaqoob P, Tzortzis G, Gibson GR (2008) Modulation of the fecal microflora profile and immune function by a novel trans-galactooligosaccharide mixture (B-GOS) in healthy elderly volunteers. Am J Clin Nutr 88:1438–1446

Vulevic J, Juric A, Tzortzis G, Gibson GR (2013) A mixture of trans-galactooligosaccharides reduces markers of metabolic syndrome and modulates the fecal microbiota and immune function of overweight adults. J Nutr 143:324–331

Walker AW, Duncan SH, McWilliam Leitch EC, Child MW, Flint HJ (2005) pH and peptide supply can radically alter bacterial populations and short-chain fatty acid ratios within microbial communities from the human colon. Appl Environ Microbiol 71:3692–3700

Walker AW, Ince J, Duncan SH, Webster LM, Holtrop G, Ze X, Brown D, Stares MD, Scott P, Bergerat A, Louis P, McIntosh F, Johnstone AM, Lobley GE, Parkhill J, Flint HJ (2011) Dominant and diet-responsive groups of bacteria within the human colonic microbiota. ISME J 5:220–230

Weaver CM, Martin BR, Nakatsu CH, Armstrong AP, Clavijo A, McCabe LD, McCabe GP, Duignan S, Schoterman MH, van den Heuvel EG (2011) Galactooligosaccharides improve mineral absorption and bone properties in growing rats through gut fermentation. J Agric Food Chem 59:6501–6510

Westerbeek EA, van den Berg A, Lafeber HN, Fetter WP, van Elburg RM (2011) The effect of enteral supplementation of a prebiotic mixture of non-human milk galacto-, fructo- and acidic oligosaccharides on intestinal permeability in preterm infants. Br J Nutr 105:268–274

Whelan K (2011) Probiotics and prebiotics in the management of irritable bowel syndrome: a review of recent clinical trials and systematic reviews. Curr Opin Clin Nutr Metab Care 14:581–587

Whelan K (2013) Mechanisms and effectiveness of prebiotics in modifying the gastrointestinal microbiota for the management of digestive disorders. Proc Nutr Soc 72:288–298

Yap KW, Mohamed S, Yazid AM, Maznah I, Meyer DM (2005) Dose-response effects of inulin on the faecal fatty acids content and mineral absorption of formula-fed infants. Nutr Food Sci 35:208–219

Yoo H-D, Kim D, Paek S-H, Oh S-E (2012) Plant cell wall polysaccharides as potential resources for the development of novel prebiotics. Biomol Ther 20:371–379

Ze X, Duncan SH, Louis P, Flint HJ (2012) *Ruminococcus bromii* is a keystone species for the degradation of resistant starch in the human colon. ISME J 6:1535–1543

Zhou J, Martin RJ, Tulley RT, Raggio AM, McCutcheon KL, Shen L, Danna SC, Tripathy S, Hegsted M, Keenan MJ (2008) Dietary resistant starch upregulates total GLP-1 and PYY in a sustained day-long manner through fermentation in rodents. Am J Physiol Endocrinol Metab 295:E1160–E1166

# How to Manipulate the Microbiota: Fecal Microbiota Transplantation

# 10

Susana Fuentes and Willem M. de Vos

**Abstract**

Fecal microbiota transplantation (FMT) is a rather straightforward therapy that manipulates the human gastrointestinal (GI) microbiota, by which a healthy donor microbiota is transferred into an existing but disturbed microbial ecosystem. This is a natural process that occurs already at birth; infants are rapidly colonized by a specific microbial community, the composition of which strongly depends on the mode of delivery and which therefore most likely originates from the mother (Palmer et al. 2007; Tannock et al. 1990). Since this early life microbial community already contains most, if not all, of the predominantly anaerobic microbes that are only found in the GI tract, it is reasonable to assume that early life colonization is the ultimate natural fecal transplantation.

**Keywords**

Gastrointestinal microbiota • Microbial ecology • Donor • *Clostridium difficile* • IBD • Regulation • Safety

## 10.1 Microbiota Transplantation: Concept and History

Fecal microbiota transplantation (FMT) is a rather straightforward therapy that manipulates the human gastrointestinal (GI) microbiota, by which a healthy donor microbiota is transferred into an existing but disturbed microbial ecosystem. This is a natural process that occurs already at birth; infants are rapidly colonized by a specific microbial community, the composition of which strongly depends on the mode of delivery and which therefore most likely originates from the mother (Palmer et al. 2007; Tannock et al. 1990). Since this early life microbial community already contains most, if not all, of the predominantly anaerobic microbes that are only found in the GI tract, it is reasonable to assume that early life colonization is the ultimate natural fecal transplantation.

The work of Eiseman, documented over 50 years ago, is considered the first study of FMT

S. Fuentes (✉) • W.M. de Vos
Laboratory of Microbiology, Wageningen University, Dreijenplein 10, 6703 Wageningen, The Netherlands
e-mail: susana.fuentes@wur.nl; willem.devos@wur.nl


A. Schwiertz (ed.), *Microbiota of the Human Body*, Advances in Experimental Medicine and Biology 902, DOI 10.1007/978-3-319-31248-4_10

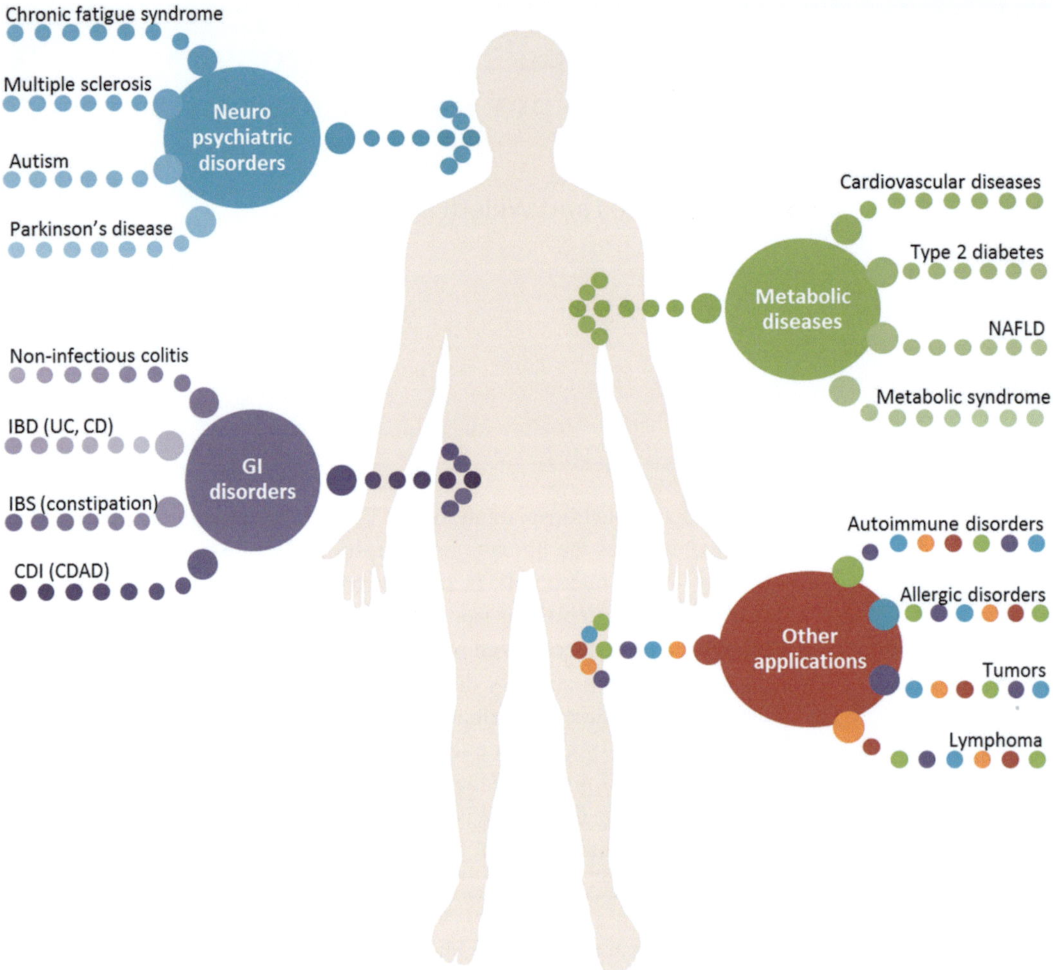


**Fig. 10.1** Applications of FMT. *GI* gastrointestinal, *IBD* inflammatory bowel disease, *UC* ulcerative colitis, *CD* Crohn's disease, *IBS* irritable bowel syndrome, *CDI C. difficile* infection, *CDAD C. difficile* associated diarrhea, *NAFLD* non-alcoholic fatty liver disease

practice in modern medicine (Eiseman et al. 1958). However, historic data indicates that this therapy dates back to the Djong-ji dynasty over 1700 years ago, and accounts of its use can be found throughout history (van Nood et al. 2014; Zhang et al. 2012; Nieuwdorp 2014; de Vos 2013). The FMT therapy has been termed in many different ways, from fecal bacteriotherapy, duodenal infusion, fecal transfusion, to even human probiotic infusion or the more recent term "rePOOPulating" (Petrof et al. 2013). Most of the published data on FMT refers to its use on recurrent *Clostridium difficile* infection (CDI) (also referred to as *C. difficile* associated diarrhea (CDAD)). FMT has, however, also been applied or is currently under investigation in many other intestinal disorders, including inflammatory bowel disease (IBD), irritable bowel syndrome (IBS) and many more (Fig. 10.1). Moreover, FMT is also being studied as potential therapy in non-intestinal conditions such as metabolic syndrome, chronic fatigue syndrome, multiple sclerosis or even certain types of cancer, as recently reviewed (Borody et al. 1989, 2011, 2012a, b; Andrews and Borody 1993; Vrieze et al. 2012; Rossen et al. 2015a; Xu et al. 2015).

Not in all cases FMT leads to successful and long-term health improvements, and variable outcomes have been reported in some diseases other than recurrent CDI. A detailed analysis of the changes in the intestinal microbiota as illustrated recently (Fuentes et al. 2014) is expected to be instrumental in further advancing the understanding of response to the therapy. Whether early diagnosed diseases, states where the microbiota is greatly disturbed or those caused by antibiotic use may respond better, are some of the hypotheses that remain to be confirmed by properly designed clinical trials.

## 10.2 Applications

### 10.2.1 *Clostridium difficile* Infection

*Clostridium difficile* infection (CDI or CDAD) is thus far, of all diseases treated by FMT, the most studied and with best outcomes. *C. difficile* is a Gram-positive and sporulating anaerobe belonging to the *Clostridium* cluster XI (phylum Firmicutes) that may produce toxins and carry one or more antibiotic-resistance determinants. CDI is an antibiotic-induced severe form of diarrhea and colitis, mainly caused by toxin-producing *C. difficile*. A stable, healthy gut microbiota can be "tipped" by certain triggers (such as antibiotic use) leading to a temporal and unstable state (Lahti et al. 2014). These critical transitions have been also found to be associated with disease-related conditions such as obesity or in the elderly. *C. difficile* overgrowth is usually reinforced by a disturbed, low diversity GI microbiota (Fig. 10.2) and, in many occasions, by reiterated antibiotic treatments.

The appearance of more virulent and antibiotic-resistant strains have led to an increase in the incidence of CDI, mostly in compromised individuals such as the elderly or patients suffering from GI disorders, but also in populations typically considered as healthy (Kelly and LaMont 2008). In view of this increasing morbidity and associated mortality there is an urgent

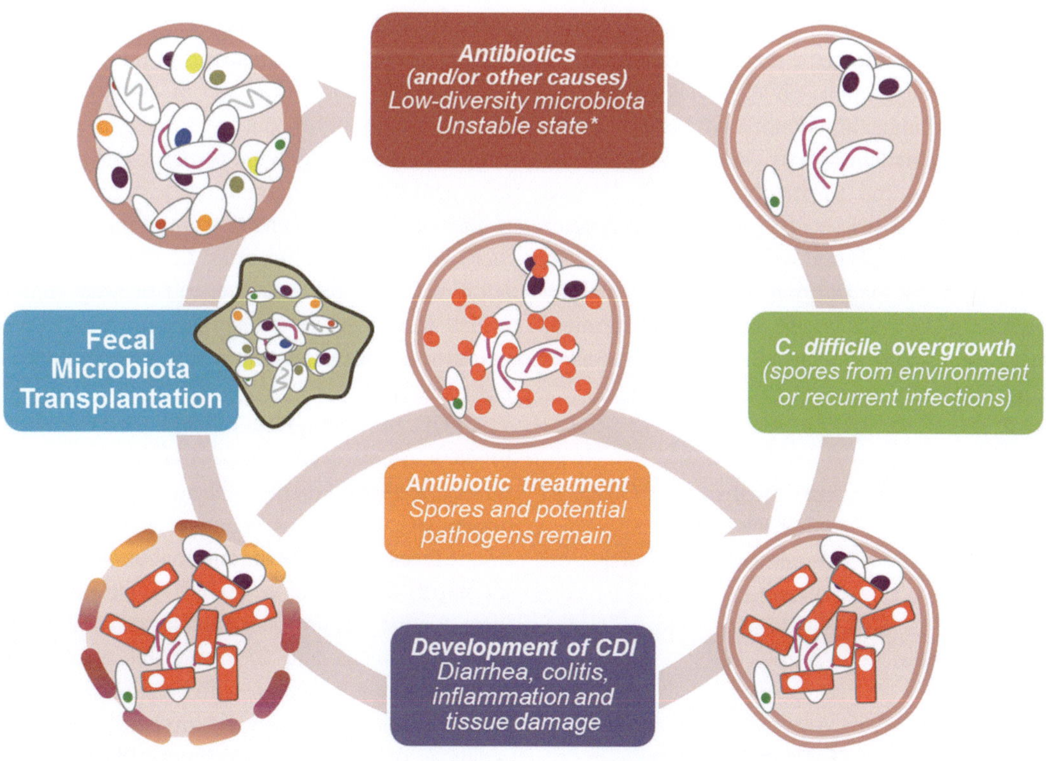


**Fig. 10.2** Cycle of *C. difficile* infection and use of FMT (*Unstable state as described by Lahti et al. (2014))

need for fast and more effective treatments. Currently, several therapies exist for the treatment of CDI, ranging from the more drastic measures such as extensive antibiotic treatments to more mild and adjuvant therapies such as probiotics (Na and Kelly 2011; Tannock et al. 2010; Lo Vecchio and Zacur 2012). However, FMT is currently the treatment with highest success rate especially for patients suffering from longstanding recurrent forms of the disease (van Nood et al. 2013) (Table 10.1).

The first randomized clinical trial of FMT on CDI was published in 2013 (van Nood et al. 2013). In this study, a total of 42 patients with recurrent CDI were randomized into three arms: FMT, Vancomycin (standard treatment for CDI) and Vancomycin plus bowel lavage (treatment routinely used prior to FMT therapy). Recovery was 81 % for the FMT arm (94 % after second infusion), 31 % for the Vancomycin arm and 23 % for the Vancomycin plus bowel lavage arm. Differences in recovery rates between the study groups was so large that the non-FMT arms of the trial were stopped for ethical reasons after an interim analysis. After the therapy, patients receiving FMT showed a raise in their GI microbiota diversity, approaching levels similar to those found in healthy donors. An increase in the relative abundance of members of the Bacteroidetes and *Clostridium* clusters IV and XIVa, and a decrease in Proteobacteria species were also found. Based on these findings, it was concluded that long-term restoration of a healthy microbiota by FMT creates a sustainable homeostatic status that can prevent future recurrences (Fig. 10.2) (Fuentes et al. 2014).

**Table 10.1** Recovery rates for (recurrent) *C. difficile* infection treatments from randomized controlled trials[a]

| Recurrences | Treatment | Recovery (%) |
|---|---|---|
| Original infection | Vancomycin | 70 |
| | Fidaxomicin | 82 |
| First recurrence | Vancomycin | 59 |
| | Fidaxomicin | 76 |
| >1 recurrences | Vancomycin | 4 |
| | FMT | 94 |

[a]Modified from Keller and Kuijper (2015)

## 10.2.2 Inflammatory Bowel Disease

Inflammatory bowel disease (IBD) is a lifelong GI disorder characterized by a series of relapses and remissions without a permanent cure. Recent reviews show that the success of FMT in IBD is very variable, and to date very limited data is available to investigate its efficacy and safety (Rossen et al. 2015a; Anderson et al. 2012). Reports of FMT in IBD, including ulcerative colitis (UC) and Crohn's disease (CD), begin in the 1980s, when researchers described a remission of UC for 6 months after FMT (Bennet and Brinkman 1989). Since then, results from randomized clinical trials in over 100 IBD patients (in some cases concurrent with CDI) have shown an improvement of symptoms and remissions that range from 6 months up to 13 years, with CD generally being less responsive to treatment than UC, and where very frequently several treatments are needed (Borody et al. 2012a, 2014; Kao et al. 2014; Greenberg et al. 2013) (Table 10.2).

In a very recent study on FMT in pediatric patients with Crohn's disease (Suskind et al. 2015), it was found via analysis of the gut microbiota that a stable transfer of healthy donor microbiota in seven out of nine patients had occurred. According to the pediatric Crohn's disease activity index, these seven patients were in remission at 2 weeks and five out of nine patients were still in remission after 12 weeks with no additional therapy. However, no or very minor effect was seen in those patients where the engraftment of the healthy donor microbiota was not successful.

In UC, remission rates of FMT vary from 0 % to 68 %. In a recent double-blind, randomized clinical trial, 37 patients with mild to moderately active UC received FMT with either healthy donor feces or their own (autologous infusion as a control) (Rossen et al. 2015b). Recovery rates were for the per protocol analysis of 41.2 % for the donor feces and 25 % for the autologous transplantation, however, with no significant difference mostly due to low numbers ($p=0.29$). Analysis of the fecal microbiota of patients 12 weeks after treatment revealed that the microbi-

**Table 10.2** Clinical outcome data on FMT in IBD

| Author | Year | N | Diagnose | Age[d] | Clinical improvement | Clinical remission | Reference |
|---|---|---|---|---|---|---|---|
| Angelberger | 2013 | 5 | Refractory UC | 27 (22–51) | 20 % | NR | Angelberger et al. (2013) |
| Borody | 2012 | 62 | Active UC | M: 42.3 ± 11.5<br>F: 48.45 ± 16.49 | 92 % | 68 % | Borody et al. (2012a) |
| Greenberg | 2013 | 16 | Refractory CD (2)<br>UC (14)[b] | 39 (20–75) | 63 % | NR | Greenberg et al. (2013) |
| Kump | 2013 | 6 | Refractory UC | 36 (17–52) | 33 % | 0 % | Kump et al. (2013) |
| Kunde | 2013 | 10[a] | Active UC[c] | 7–20 | 70 % | 30 % | Kunde et al. (2013) |
| Vermeire | 2012 | 4 | Refractory CD | 37.5 (29–50) | 0 % | 0 % | Vermeire et al. (2012) |
| Rossen | 2015 | 48 | Active UC | 30–56 | NR | 30.4–41.2 %[e] | Rossen et al. (2015b) |

Modified from Rossen et al. (2015a)
*NR* not reported
[a]1 subject did not tolerate treatment, this subject was considered to be a treatment failure. Endpoint data of the study was adjusted in this table
[b]Concurrent CDI in four UC patients
[c]Active disease diagnosed by colonoscopy <6 months before the enrolment
[d]Mean ± SD or median, range/IQR
[e]For "intention to treat" and "per protocol" analysis respectively

ota of patients responding to the therapy in the FMT group was similar to that of their healthy donors. Remission could also be associated with increased proportions of *Clostridium* clusters IV and XIVa, similar to what has been observed in CDI.

These results highlight one of the limitations of FMT that lies in the importance of case by case targeted therapies rather than undefined mixtures of bacteria. There are currently several clinical trials designed to investigate FMT on IBD. In clinicaltrials.gov, open studies in recruiting phase for FMT in IBD include six in UC (one of them in UC associated pouchitis), two on CD and four on IBD in general. Results from these trials will in time shed some more light on this issue.

### 10.2.3 Other Applications

Potential applications of FMT are unlimited, as the GI microbiota is known to play an important role in numerous physiological processes. As previously mentioned FMT is under investigation for many (non-) intestinal disorders, ranging from neurological disorders to metabolic syndromes (Fig. 10.1).

In some cases, colitis can develop not only as a complication of *C. difficile* overgrowth, but as a result of other triggers such as antibiotic use. FMT has proven to be of help for these cases as well. A recent case study reported the follow up of a 46-year-old man for over a year (Satokari et al. 2014). Contrary to CDI, symptoms were not accompanied by an apparently disturbed fecal microbiota prior to the therapy; the patient showed high Bacteroidetes levels but within the normal range observed in healthy individuals. These levels were further increased 2 days after FMT, but the microbiota composition quickly shifted to one that was more dominated by Firmicutes 2 weeks after infusion. Microbiota analysis revealed an increased bacterial diversity in the rectal mucosa and a stable fecal microbiota up to 3 months after treatment. In this case of non-infectious colitis FMT was able to resolve the symptoms and restore normal GI-function, possibly by acting against a persistent low-grade inflammation.

FMT therapy has also been used in cases of irritable bowel syndrome (IBS), as well as chronic constipation (concomitant to IBS) where up to 76% of the treated patients reported improvement of symptoms including abdominal pain or bloating, among others, with a follow up period of up to a year (Andrews and Borody 1993; Pinn et al. 2013). Positive results were also reported in cases of constipation associated to UC (Rossen et al. 2015a).

In addition, FMT has an important therapeutic potential in non-intestinal disorders. The intestinal microbiota has been shown to play an important role in local and systemic inflammation, key factor for obesity and insulin resistance (Verdam et al. 2013; Turnbaugh et al. 2006; Udayappan et al. 2014). A few reported studies evaluated the effect of manipulating the microbiota by means of fecal transplantation in diseases associated to obesity, such as type 2 diabetes or metabolic syndrome (Vrieze et al. 2012, 2014; Hartstra et al. 2015). In a recent study on a cohort of 18 patients diagnosed with metabolic syndrome, a significant increase was observed in insulin sensitivity after reception of healthy donor feces as compared to the autologous placebo group (i.e. those patients who received their own fecal material) (Vrieze et al. 2012). These patients showed an enrichment of butyrate-producing species (able to produce butyric acid from lactate and acetate) such as *Roseburia intestinalis* or *Eubacterium hallii*. Butyrate is a short chain fatty acid of great importance in colonic function, and known to be related to obesity and pain sensation (Duncan et al. 2007; Hamer et al. 2008; Louis and Flint 2009; Vanhoutvin et al. 2009). Butyrogenic bacteria have been shown to be depleted from patients suffering from CDI (Antharam et al. 2013).

Therapies including neurological diseases such as Parkinson's disease or multiple sclerosis are also being investigated (Borody et al. 2011; Borody and Khoruts 2011; Ananthaswamy 2004). Anecdotal observations indicated that after treating CDI patients who suffered from Parkinson's disease by FMT, symptoms of Parkinson's were reduced (Borody et al. 2013). Autoimmune disorders, certain tumours or even lymphoma are some of the additional applications that have been investigated for treatment with FMT (Xu et al. 2015).

## 10.3 Methodology and Donor Selection

Many different methodologies for FMT have been published (Gough et al. 2011; van Nood et al. 2009; Owens et al. 2013). The route of infusion can vary from retention enema, colonoscope and nasoduodenal tube, and results of these (so far only compared in the treatment of CDI) are comparable. In most of the cases, the patient's GI tract is prepared by bowel lavage or use of laxatives. Cleaning of the bowel has been shown to have an impact on the GI microbiota composition (Jalanka et al. 2014). In a recent study, 23 healthy individuals were given a bowel lavage either in two separate doses of 1 l or in a single 2 l dose. This treatment had an important effect on the microbiota, by loss of both microbial load and diversity as well as subject specific composition. Cleaning by means of a double dose had a less disturbing effect on the microbiota than the single dose, and it was associated with increased levels of Proteobacteria among others.

Stool samples are screened for various parasites and bacterial pathogens. A shortened version of donor and patient screening as well as therapy protocol for FMT (for CDI) is shown in Table 10.3. More detailed protocols and questionnaires can be found in van Nood et al. (supplementary material) (van Nood et al. 2013).

Donor selection has been hypothesized to be essential for success of FMT, although key characteristics are still unknown. Increasing evidence shows that there is a need for donor-patient matching, and that some donors work better than others (Fuentes et al. 2014; Borody et al. 2014) (Rossen et al. 2015b). As mentioned above, donor selection criteria are based on extensive questionnaires that lead to exclusion following thorough examination of (non-) GI diseases as well as other parameters. Blood samples are also screened for numerous antibodies against potentially transmittable diseases, such as hepatitis (A/B/C), HIV or parasites. If multiple samples are required from selected donors, questionnaires and screenings are frequently repeated. Donors that have led to better outcomes in previous FMT

**Table 10.3** Shortened Amsterdam protocol for FMT for CDI

| Donor | Patient |
|---|---|
| 1. Screening | 1. Selection criteria for FMT |
| Questionnaire for risk factors and potentially transmittable disorders | >2 CDI recurrences |
| | Severe therapy resistant CDI |
| Exclusion criteria: | |
| Antibiotic use (<3 months) | |
| GI symptoms (diarrhea, constipation or IBS symptoms) | |
| Recent travel to areas of endemic GI pathogens | |
| IBD | |
| GI malignancies or polyposis | |
| 2. Laboratory tests: | 2. Pre-treatment of patient |
| Blood tests/serology | Vancomycin treatment >4 days (stopped 1 day before FMT) |
| Feces | Bowel lavage (day before FMT) |
| | Placement of nasoduodenal tube (day of FMT) |
| 3. 1 day before donation | |
| Questionnaire for recent health related issues | |
| **FMT** | |
| Preparation of donor feces: | |
| Dilution of feces in sterile saline | |
| Filter through unfolded gauzes and collect in a closed bottle | |
| Infusion: | |
| Infuse through nasoduodenal tube the donor feces solution | |
| Flush the nasoduodenal tube with water | |
| Remove tube 30 min after infusion | |
| Offer lemonade to patient after removal | |

More information in van Nood et al. (2013) and Keller and Kuijper (2015)

trials could be preselected as high efficiency donors, as well as those with higher levels of the already mentioned butyrate producing bacteria or health related strains such as *Faecalibacterium prausnitzii* or *Akkermansia muciniphila* (Sokol et al. 2008; Everard et al. 2013; Kang et al. 2013; Berry and Reinisch 2013).

## 10.4 Synthetic Communities

Besides infusion of diluted and filtered feces, other preparations have also been used as "fecal material", including frozen preparations or selections of several bacterial species grown in vitro, most of them isolated from fresh fecal samples (Petrof et al. 2013; Hamilton et al. 2013; Youngster et al. 2014; Tvede and Rask-Madsen 1989). These preparations have been shown to provide similar outcomes as freshly prepared stools without additional side effects, becoming a practical alternative to FMT (Satokari et al. 2015; Tvede et al. 2015). Current research aims to identify specific communities associated with different diseases (as well as a defined "healthy" microbiota). This would allow for targeted treatment therapies for specific diseases with well-characterized strains or active ingredients that are now administered "blindly" by fecal transplantation. To this end, several public and private sectors are investigating defined synthetic communities. These communities can be controlled and reproduced as regular treatments without the need of human donors, and under more controlled conditions.

One of the first studies involving a defined mixture of strains as treatment for recurrent CDI was published in The Lancet in 1989 (Tvede and Rask-Madsen 1989). A group of six patients was treated by an enema with either stool samples or with a mixture of ten cultured bacteria (*Clostridium innocuum*, *C. ramosum*, *C. bifermentans*, *Bacteroides ovatus*, *B. vulgatus*, *B. thetaiotaomicron*, *Peptostreptococcus productus*, *Streptococcus faecalis*, and two strains of *Escherichia coli*). After treatment, the response was rapid, and all patients were asymptomatic and showed no *C. difficile* toxin after 24 h. Interestingly one of the patients responded better to the mixture than to the FMT. It was not until 2013 that a mixture of bacterial strains was used again for the treatment of recurrent CDI in two patients, in this case containing 33 strains isolated from one healthy donor (Petrof et al. 2013), coining the term already mentioned, "rePOOPulating" of the gut.

## 10.5 Regulations and Safety

There is an increasingly pressing need to regulate fecal transplants due to the remarkable interest that this therapy generates, with publications not properly documented or even videos online for self-infusion. In 2013 the US Food and Drug Administration (FDA) classified human feces (for medical use) as a drug, in hopes of regulating and standardizing its use. This implied the requirement to submit an IND (investigational new drug) application when using this therapy. However, after physicians, researchers and patients raised their concern about access to this treatment becoming growingly constrained, the FDA decided to not enforce the IND for its use on recurrent CDI. Top physicians in the field indicated that reclassifying human feces as a tissue product or a different category (such as blood) would facilitate it's application while maintaining safety regulations (Smith et al. 2014).

While in the UK fecal transplants are accepted as therapy for CDI, where it is considered safe and effective, in the US current regulations are somewhat stricter and require that donors (and their stools) must be known and screened by the physicians involved in the treatment. This rule led to the closing of stool banks such as OpenBiome, a non-profit organization from the Massachusetts Institute of Technology. OpenBiome carried out 17 blood and stool screening assays and supplied samples to several US hospitals (more than 100 treatments in its first 3 months) (Smith et al. 2014). Stool banks, similarly as blood banks do, could be a source for thoroughly screened donor material. Not only would this have an impact on the safety of the procedure, avoiding desperate, risky and largely uncontrolled treatments at home, but also it would make it cheaper and more accessible. However, as of 2015 the FDA allows for FMT for CDI, while for other applications an IND is still mandatory (Smith et al. 2014). In order for stool banks to succeed, the FDA would need to either reclassify FMT as suggested by experts, or approve its use in disorders different from CDI without the burden of bureaucracy.Very recently, the Netherlands Donor Feces Bank (NDFB) was initiated to enable safe and cost-effective FMT for patients with recurrent CDI, and to support further research.

Among the limitations of FMT one can list the observation that for some therapies (e. g. in patients suffering from IBD) several repeated FMT applications are necessary (Ratner 2014). This will make the procedure rather costly as compared to standard treatments, and as the effect is variable, not an attractive alternative. This is not the case for CDI, where cost-effectiveness has been evaluated, and FMT was estimated to save costs when compared to standard antibiotherapy (van Nood 2015).

Although the main advantage of FMT is that it has limited adverse effects, these are still an issue. Although the most frequent are mild adverse effects often self-limiting, especially those related to abdominal discomfort (such as pain, bloating, nausea or even vomiting), severe adverse effects have also been reported including, among others, intestinal perforation, post-transplant sepsis or bacteraemia, although in many cases causal relation to the therapy could not be established (Rossen et al. 2015a). In several of the FMTs, a careful analysis of the patients health status has been reported, which included the observation that a transient and small increase in CRP levels was observed after 2–3 days both for the autologous (own microbiota) as healthy donor transplantation (Vrieze 2013).

Furthermore, only few studies include long term follow up of patients. Long-term consequences, although yet unreported and mostly unknown, are still a risk. Studies that would allow evaluating the potential risk of FMT with respect to possible future infections, auto-immune or metabolic diseases or even cancer are still required. At present, the main fear is that fecal material can carry viruses or other infectious diseases, although to date there are no cases of transmission of infectious diseases directly related to FMT. Additionally,

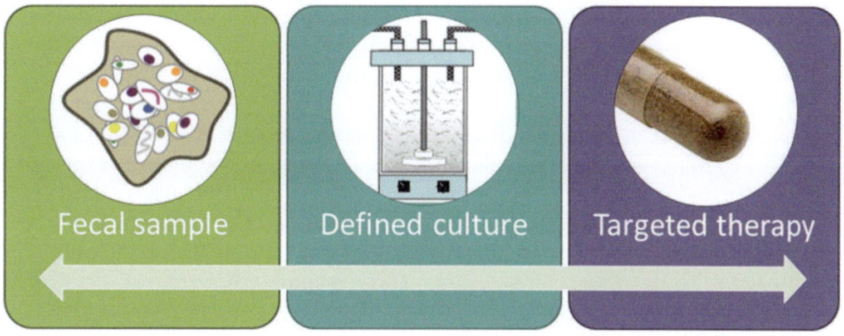


**Fig. 10.3** Evolution of FMT therapy

changes in an individual's GI microbiota composition could lead to a shift in phenotype to one of higher susceptibility for microbiota related diseases, such as obesity and other conditions. A recent case study reports a successfully treated patient who 16 months after therapy developed new-onset obesity, after receiving FMT from an overweight donor (Alang and Kelly 2015).

## 10.6 Microbiota Transplantation: The future

Research in the field of fecal transplantation is rapidly growing, with countless applications. Results from the last decade are showing a rapid evolution of how we will approach the application of this therapy (Fig. 10.3). Whether the future of FMT will be in the form of diluted healthy donor feces, or defined consortia of cultured species in a pill, remains to be assessed.

Identification of essential microbes (and/or microbial functions) for specific diseases will also be of high importance, as it has been shown that FMT is not "one size fits all" (Rossen et al. 2015b). Identifying the smallest number of strains (also known as the minimal microbiome (de Vos 2013)) necessary to exert a beneficial effect on the host by establishing a stable community, is still a major challenge. To this end, interactions with the host at a systems level will also be of high importance (Martins dos Santos et al. 2010). Identification of the minimal microbiome is not only important from a mechanistic point of view, but also in terms of practicality. The amounts of fecal material that are currently used for infusion can vary from 30 g (or cc) to up to 500 g (Rossen et al. 2015a). This would prove challenging when wanting to translate an infusion of fecal material into a pill format.

Up to date, evidence shows that FMT is an efficient, simple therapy with rather positive outcomes. Modulation of the gut microbiota by FMT can lead to the establishment of a stable microbiota composition that tends to remain in a healthy state of homeostasis. However, results from well controlled, randomized, double blind clinical trials are urgently needed, and will be key to fine tune this therapy.

**Acknowledgements** We would like to thank all our collaborators, especially those involved in the mentioned clinical trials. This work was supported by CVON-IN CONTROL (Cardiovascular research, The Netherlands, www.cvon.eu), Spinoza Awards and Gravitation Programmes from the Netherlands Organisation for Scientific Research (NWO).

## References

Alang N, Kelly CR (2015) Weight gain after fecal microbiota transplantation. Open Forum Infect Dis 2(1):ofv004. doi:10.1093/ofid/ofv004

Ananthaswamy A (2011) Faecal transplant eases symptoms of Parkinson's disease. New Sci 209(2796):8–9

Anderson JL, Edney RJ, Whelan K (2012) Systematic review: faecal microbiota transplantation in the management of inflammatory bowel disease. Aliment Pharmacol Ther 36(6):503–516

Andrews PJ, Borody TJ (1993) "Putting back the bugs": bacterial treatment relieves chronic constipation and symptoms of irritable bowel syndrome. Med J Aust 159(9):633–634

Angelberger S et al (2013) Temporal bacterial community dynamics vary among ulcerative colitis patients after fecal microbiota transplantation. Am J Gastroenterol 108(10):1620–1630

Antharam VC et al (2013) Intestinal dysbiosis and depletion of butyrogenic bacteria in Clostridium difficile infection and nosocomial diarrhea. J Clin Microbiol 51(9):2884–2892

Bennet JD, Brinkman M (1989) Treatment of ulcerative colitis by implantation of normal colonic flora. Lancet 1(8630):164

Berry D, Reinisch W (2013) Intestinal microbiota: a source of novel biomarkers in inflammatory bowel diseases? Best Pract Res Clin Gastroenterol 27(1):47–58

Borody TJ, Khoruts A (2011) Fecal microbiota transplantation and emerging applications. Nature reviews. Gastroenterol Hepatol 9:99–96

Borody TJ et al (1989) Bowel-flora alteration: a potential cure for inflammatory bowel disease and irritable bowel syndrome? Med J Aust 150(10):604

Borody T et al (2011) Fecal microbiota transplantation (FMT) in multiple sclerosis (MS). Am J Gastroenterol 106:S352–S352

Borody T et al (2012a) Fecal microbiota transplantation in ulcerative colitis: review of 24 years experience. Am J Gastroenterol 107:S665–S665

Borody T et al (2012b) Bacteriotherapy in chronic fatigue syndrome (CFS): a retrospective review. Am J Gastroenterol 107:S591–S592

Borody TJ, Paramsothy S, Agrawal G (2013) Fecal microbiota transplantation: indications, methods, evidence, and future directions. Curr Gastroenterol Rep 15(8):337

Borody TJ, Finlayson S, Paramsothy S (2014) Is Crohn's disease ready for fecal microbiota transplantation? J Clin Gastroenterol 48(7):582–583

de Vos WM (2013) Fame and future of faecal transplantations – developing next-generation therapies with synthetic microbiomes. J Microbial Biotechnol 6(4):316–325

Duncan SH et al (2007) Reduced dietary intake of carbohydrates by obese subjects results in decreased concentrations of butyrate and butyrate-producing bacteria in feces. Appl Environ Microbiol 73(4):1073–1078

Eiseman B et al (1958) Fecal enema as an adjunct in the treatment of pseudomembranous enterocolitis. Surgery 44(5):854–859

Everard A et al (2013) Cross-talk between Akkermansia muciniphila and intestinal epithelium controls diet-induced obesity. Proc Natl Acad Sci U S A 110(22):9066–9071

Fuentes S et al (2014) Reset of a critically disturbed microbial ecosystem: faecal transplant in recurrent Clostridium difficile infection. ISME J 8:1621–1633

Gough E, Shaikh H, Manges AR (2011) Systematic review of intestinal microbiota transplantation (fecal bacteriotherapy) for recurrent Clostridium difficile infection. Clin Infect Dis 53(10):994–1002

Greenberg A et al (2013) Long-term follow-up study of fecal microbiota transplantation (FMT) for inflammatory bowel disease (IBD). Am J Gastroenterol 108:S540–S540

Hamer HM et al (2008) Review article: the role of butyrate on colonic function. Aliment Pharmacol Ther 27(2):104–119

Hamilton MJ et al (2013) High-throughput DNA sequence analysis reveals stable engraftment of gut microbiota following transplantation of previously frozen fecal bacteria. Gut Microbes 4(2):125–135

Hartstra AV et al (2015) Insights into the role of the microbiome in obesity and type 2 diabetes. Diabetes Care 38(1):159–165

Jalanka J et al (2014) Effects of bowel cleansing on the intestinal microbiota. Gut 64:1562–1568

Kang CS et al (2013) Extracellular vesicles derived from gut microbiota, especially Akkermansia muciniphila, protect the progression of dextran sulfate sodium-induced colitis. PLoS One 8(10):e76520

Kao D et al (2014) Fecal microbiota transplantation inducing remission in Crohn's colitis and the associated changes in fecal microbial profile. J Clin Gastroenterol 48(7):625–628

Keller JJ, Kuijper EJ (2015) Treatment of recurrent and severe Clostridium difficile infection. Annu Rev Med 66:373–386

Kelly CP, LaMont JT (2008) Clostridium difficile – more difficult than ever. N Engl J Med 359(18):1932–1940

Kump PK et al (2013) Alteration of intestinal dysbiosis by fecal microbiota transplantation does not induce remission in patients with chronic active ulcerative colitis. Inflamm Bowel Dis 19(10):2155–2165

Kunde S et al (2013) Safety, tolerability, and clinical response after fecal transplantation in children and young adults with ulcerative colitis. J Pediatr Gastroenterol Nutr 56(6):597–601

Lahti L et al (2014) Tipping elements in the human intestinal ecosystem. Nat Commun 5:4344

Lo Vecchio A, Zacur GM (2012) Clostridium difficile infection: an update on epidemiology, risk factors, and therapeutic options. Curr Opin Gastroenterol 28(1):1–9

Louis P, Flint HJ (2009) Diversity, metabolism and microbial ecology of butyrate-producing bacteria from the human large intestine. FEMS Microbiol Lett 294(1):1–8

Martins dos Santos V, Muller M, de Vos WM (2010) Systems biology of the gut: the interplay of food, microbiota and host at the mucosal interface. Curr Opin Biotechnol 21(4):539–550

Na X, Kelly C (2011) Probiotics in clostridium difficile infection. J Clin Gastroenterol 45(Suppl): S154–S158

Nieuwdorp M (2014) Faecal microbiota transplantation. Br J Surg 101(8):887–888

Owens C, Broussard E, Surawicz C (2013) Fecal microbiota transplantation and donor standardization. Trends Microbiol 21(9):443–445

Palmer C et al (2007) Development of the human infant intestinal microbiota. PLoS Biol 5(7):e177

Petrof EO et al (2013) Stool substitute transplant therapy for the eradication of Clostridium difficile infection: 'RePOOPulating' the gut. Microbiome 1(1):3

Pinn D, Aroniadis O, Brandt L (2013) Follow-up study of fecal microbiota transplantation (FMT) for the treatment of refractory irritable bowel syndrome (IBS). Am J Gastroenterol 108:S563–S563

Ratner M (2014) Fecal transplantation poses dilemma for FDA. Nat Biotechnol 32(5):401–402

Rossen NG et al (2015a) Faecal microbiota transplantation as novel therapy in gastroenterology: a systematic review. World J Gastroenterol: WJG 9:342–348

Rossen NG, Fuentes S, van der Spek MJ, Tijssen JG, Hartman JH, Duflou A, Lowenberg M, van den Brink GR, Mathus-Vliegen EM, de Vos WM, Zoetendal EG, D'Haens GR, Ponsioen CY (2015b) Findings from a randomized controlled trial of fecal transplantation for patients with ulcerative colitis. Gastroenterology 149:110–118 e4

Satokari R, et al. (2014) *Case Report: Fecal transplantation treatment of antibiotic-induced, noninfectious colitis and long-term microbiota follow-up.* Case Rep Med, 2014 (Article ID 913867): p. 7

Satokari R et al (2015) Simple faecal preparation and efficacy of frozen inoculum in faecal microbiota transplantation for recurrent Clostridium difficile infection – an observational cohort study. Aliment Pharmacol Ther 41(1):46–53

Smith MB, Kelly C, Alm EJ (2014) Policy: how to regulate faecal transplants. Nature 506(7488):290–291

Sokol H et al (2008) Faecalibacterium prausnitzii is an anti-inflammatory commensal bacterium identified by gut microbiota analysis of Crohn disease patients. Proc Natl Acad Sci U S A 105(43):16731–16736

Suskind DL et al (2015) Fecal microbial transplant effect on clinical outcomes and fecal microbiome in active Crohn's disease. Inflamm Bowel Dis 3:556–563

Tannock GW et al (1990) Plasmid profiling of members of the family Enterobacteriaceae, lactobacilli, and bifidobacteria to study the transmission of bacteria from mother to infant. J Clin Microbiol 28(6):1225–1228

Tannock GW et al (2010) A new macrocyclic antibiotic, fidaxomicin (OPT-80), causes less alteration to the bowel microbiota of Clostridium difficile-infected patients than does vancomycin. Microbiology 156(Pt 11):3354–3359

Turnbaugh PJ et al (2006) An obesity-associated gut microbiome with increased capacity for energy harvest. Nature 444(7122):1027–1031

Tvede M, Rask-Madsen J (1989) Bacteriotherapy for chronic relapsing Clostridium difficile diarrhoea in six patients. Lancet 1(8648):1156–1160

Tvede M, Tinggaard M, Helms M (2015) Rectal bacteriotherapy for recurrent Clostridium difficile-associated diarrhoea: results from a case series of 55 patients in Denmark 2000–2012. Clin Microbiol Infect 21(1):48–53

Udayappan SD et al (2014) Intestinal microbiota and faecal transplantation as treatment modality for insulin resistance and type 2 diabetes mellitus. Clin Exp Immunol 177(1):24–29

Van Nood E (2015) Fecal microbiota transplantation. Clinical and experimental studies, in Department of Internal Medicine. Ph.D. Thesis, Amsterdam, Academic Medical Center

van Nood E, Speelman P, Kuijper EJ, Keller JJ (2009) Struggling with recurrent Clostridium difficile infections: is donor faeces the solution? Euro Surveill 14(34):pii: 19316

van Nood E et al (2013) Duodenal infusion of donor feces for recurrent Clostridium difficile. N Engl J Med 368(5):407–415

van Nood E et al (2014) Fecal microbiota transplantation: facts and controversies. Curr Opin Gastroenterol 30(1):34–39

Vanhoutvin SA et al (2009) The effects of butyrate enemas on visceral perception in healthy volunteers. Neurogastroenterol Motility: Off J Eur Gastrointest Motility Soc 21(9):952-e76

Verdam FJ et al (2013) Human intestinal microbiota composition is associated with local and systemic inflammation in obesity. Obesity 21:E607–E615

Vermeire S et al (2012) Pilot study on the safety and efficacy of faecal microbiota transplantation in refractory Crohn's disease. Gastroenterology 142(5):S360–S360

Vrieze A (2013) The role of gut microbiota in human metabolism, in Department of Internal Medicine, Ph.D. Thesis. Amsterdam, Academic Medical Center

Vrieze A et al (2012) Transfer of intestinal microbiota from lean donors increases insulin sensitivity in individuals with metabolic syndrome. Gastroenterology 143:913–916

Vrieze A et al (2014) Impact of oral vancomycin on gut microbiota, bile acid metabolism, and insulin sensitivity. J Hepatol 60(4):824–831

Xu MQ et al (2015) Fecal microbiota transplantation broadening its application beyond intestinal disorders. World J Gastroenterol: WJG 21(1):102–111

Youngster I et al (2014) Oral, capsulized, frozen fecal microbiota transplantation for relapsing Clostridium difficile infection. JAMA 312(17):1772–1778

Zhang F et al (2012) Should we standardize the 1,700-year-old fecal microbiota transplantation? Am J Gastroenterol 107(11):1755

# Index




A. Schwiertz (ed.), *Microbiota of the Human Body*, Advances in Experimental Medicine and Biology 902, DOI 10.1007/978-3-319-31248-4

MIX
Papier aus verantwortungsvollen Quellen
Paper from responsible sources
FSC® C105338

If you have any concerns about our products,
you can contact us on
**ProductSafety@springernature.com**

In case Publisher is established outside the EU,
the EU authorized representative is:
**Springer Nature Customer Service Center GmbH**
**Europaplatz 3, 69115 Heidelberg, Germany**

Printed by Libri Plureos GmbH
in Hamburg, Germany